1000 *by* TIPS
100 ECO ARCHITECTS

1000 TIPS *by* TIPS
100 ECO ARCHITECTS

Marta Serrats, Editor

FIREFLY BOOKS

A FIREFLY BOOK

Published by Firefly Books Ltd. 2012

First printing

Publisher Cataloging-in-Publication Data (U.S.)

A CIP record for this title is available from the Library of Congress

Library and Archives Canada Cataloguing in Publication

A CIP record for this title is available from Library and Archives Canada

Published in the United States by
Firefly Books (U.S.) Inc.
P.O. Box 1338, Ellicott Station
Buffalo, New York 14205

Published in Canada by
Firefly Books Ltd.
66 Leek Crescent
Richmond Hill, Ontario L4B 1H1

Cover design: Erin R. Holmes/Soplari Design

Printed in China

This book was developed by:
LOFT Publications, S.L.
Via Laietana 32, 4º, of. 92
08003 Barcelona, España
Tel.: +34 93 268 80 88

Editorial coordinator: Aitana Lleonart Triquell
Editor: Marta Serrats
Art director: Mireia Casanovas Soley
Design and layout coordination: Claudia Martínez Alonso
Layout: Cristina Simó Perales
Cover layout: María Eugenia Castell Carballo
Translation: Cillero & de Motta

Contents

Introduction

First of all, we would like to highlight what has been the main intention of this book, which is to give the leading role to the participating architects. The book aims to reveal the trends in sustainable architecture in the most practical manner possible from architects who have worked on the final content and who have contributed, through their ideas, to the basics of sustainable architecture.

We should mention the architects often had difficulties being involved in the preparation of the content, and many refused to participate due to lack of time and workload, even though the field of architecture has been one of the sectors hardest hit by the current construction crisis.

For several decades now the concepts of ecology and sustainability have generated highly influential currents of opinion in our society, invading all areas of our life. Today everyone is working to combat climate change and address the ever more pressing need to reduce greenhouse gas emissions. In the field of architecture, significant changes and developments are evolving. However, research shows that we still have a way to go. The percentage of sustainable buildings is still very low and in open conflict with the increase in the number of building projects around the world. In the field of residential architecture alone, there is now an increase in the number of homes as a result of the growth in population. This in turn leads to greater consumption of raw materials and rising environmental costs for their transportation.

Much more participation is still required on the part of everyone involved, from architects and building contractors to manufacturers of materials and end customers, if what we hope to achieve is a world constructed of quality infrastructure that is also environmentally friendly and sustainable at the same time.

The 100 architects involved in this book present the main guidelines for responsible architecture. Among the participating architects, some have presented a brief explanation of the basics of eco-construction and green criteria applied in their work, while others have emphasized basic concepts of building a healthy home with green, renewable materials, heated with solar and geothermal energy and naturally lit. Some have opted to highlight, based on personal experience, the efficiency of using new materials or strategies to achieve a self-sufficient architecture.

We appreciate the generosity of these 100 architects who, by sharing their expertise, offer us a brief introduction to the most common terms of ecological architecture. We would like to thank them all.

Marta Serrats

0 to 1

4 Lexington Avenue, #2M
New York, New York 10010, USA
Tel.: +1 646 279 9387
www.0-to-1.com

0001 ▼
Artist retreat
Create an efficient thermal envelope,
while providing optimal natural light.
The highly insulated envelope offsets
the harsh winters. The triple-paned
north-facing windows provide soft
natural lighting for the artist to work.

0002 ▲
Street vendor tent
Modular design. A single form can be
arranged in various configurations
to meet the needs of the vendor.

0003 ▼
Street vendor tent
Explore alternative materials,
such as ice, which will melt away
when it is no longer required.

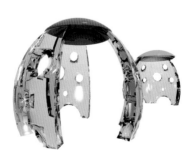

◄ 0004
Artist retreat
Efficiently designed spaces require
less energy. This single space
ascends from outdoor terrace to
work space to a small utility cube,
which doubles as a sleeping loft.

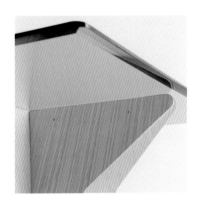

0005
Star table
Geometry allows the use of thin material. Four lightweight, ¼-inch (0.6 cm) thick interlocking planes create a strong stable base.

0006 ➤
Star table
A rapidly renewable material, bamboo plywood planes also pack flat to minimize shipping size.

◄ 0007
Townhouse renovation glass
Meeting contemporary needs while honoring existing architecture, glass walls merge with existing historic moldings. Natural light permeates to spaces without windows, reducing the energy needs for artificial lighting.

◄ 0008
Townhouse renovation stair
Balancing design aesthetics with embodied energy, a transparent stair was engineered to minimize the materials and maximize its strength.

0010 ▲
W House exterior
Form is inspired by the natural site. A three-story split-level house follows the slope of the site. During warm weather, the three-story stair space allows air to naturally move through the house, from low to high, providing natural cooling and ventilation.

0009 ➤
W House interior
A hearth rethought. At the center of the house are a fireplace and stair. The fireplace touches all spaces, providing a gathering place, acting as structural support, and distributing radiant heat to all the floors.

Amunt Architekten Martenson und Nagel Theissen

Schervierstrasse 66
Aachen, 52066 Germany
Tel.: +49 711 8496341
www.amunt.info

0011 ➤
Thinking from scratch helps to break new ground and question standards.
Instead of reducing the design merely to one function, one condition or one task, JustK is the result of multiple design aspects that were amalgamated into the architecture of the building.

0012 ▼
Use small, simple, low-tech and low-energy solutions.
The staggered main living space creates various room heights, which separates the ground floor and automatically creates different climate zones in the winter. Cold outside air remains trapped in the lower entrance area while moderate temperatures prevail in the kitchen-living room, and the sitting room higher up is warmest.

◄ **0013**
Optimize rather than maximize
Clever overlapping means that the ceiling of the bathroom acts as a sleeping loft for the neighboring children's bedroom.

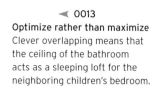

◀ 0014
Use and value renewable resources and services
Wood was the main material used consistently throughout the structure and interior surfaces; as a renewable raw material, it was chosen not least because of its favorable energy balance.

0015 ▼
integrate and evolve the existing structures
The dark red clinker facade of the original building interlocks with the red-brown unplastered pumice and light-weight concrete bricks of the extension. It interprets the existing material in a contemporary manner and combines with the original color to form a new compact building volume. 1+1=1.

0016 ▽

Using spaces appropriately
and making the most of them
creates a timeless building
The top-floor living-room hall
provides room for retreat and serves
as a reading and television area.
If splitting up the house, there is
an optional build-in kitchen.

0017 ➤

**Use and value diversity,
diversity creates potentials**
In the warmer months, this living space
can be extended by the 129-sq.-foot
(12 m²) balcony and the 248-sq.-
foot (23 m²) forecourt. Living space
is everywhere in a good home.

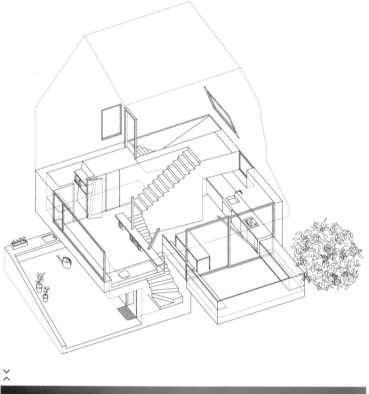

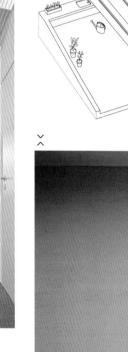

0018 ▼

Creative use of space and response to change support sustainability and flexibility. This house can be split into two living units with separate entrances without much effort should the family situation change. The total area of the house is 1,485 sq. feet (138 m²); one unit would measure 872 sq. feet (81 m²) and the other 613 sq. feet (57 m²).

◄ **0019**

Activate existing potential

Extending the house toward the garden, opening up the staircase to the upper floor, and the new wall panelling of the stairs with integrated handrail, have transformed the role of the old wooden staircase.

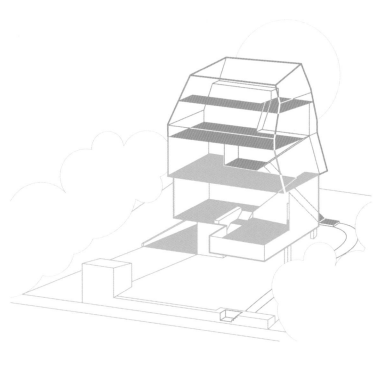

0020 ▼

Building is not always the answer.

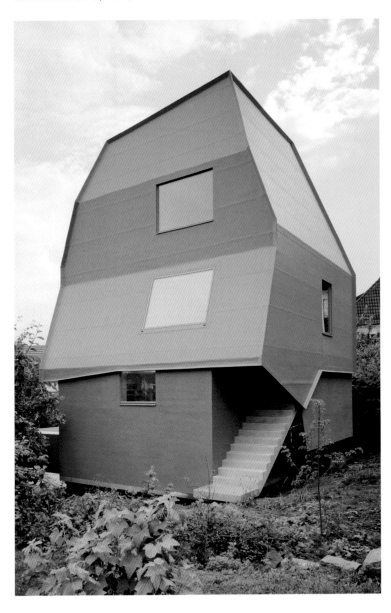

Andrea Tognon Architecture

14 Via Pericle
Milano, 20126, Italy
Tel.: +39 348 5809677
www.atognon.com

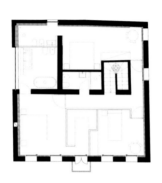 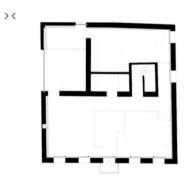 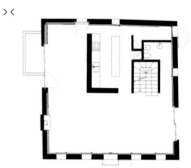 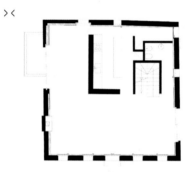

0021 ▲
The roof was a concrete slab jutting out in a very inelegant and badly proportionated way, so we decided to add the corner that was missing to complete the square floor plan.

0022 ➤
Called Residence O, the project also involved replacing the existing overhanging roof with one that sits flush with the edges of the exterior walls and a complete redesign of the interior.

◄ 0023

0023
The building we were called to refurbish was built in the 1970s as an imitation of the vernacular architecture of the Veneto countryside. It was looking pretty fake. The floor plan was a square where a corner was missing (so it was an L shape).

0024 ▲
The entire interior layout was redesigned, all the walls and roof were insulated and the heating system was switched to solar energy.

0025 ➤

Because we totally redefined the insulation parameter, we reshaped the profile of the building, cutting the old roof edge and redesigning the junction between roof and perimeter walls.

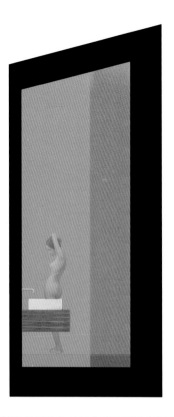

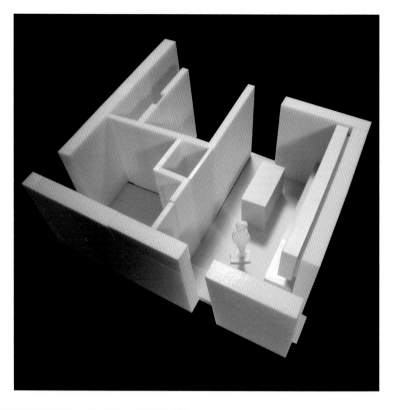

0026 ⏶

New wall and ceiling panels increase the thermal insulation of the house, reducing the need for heating, which is from solar power.

0027 ⏶

All of the walls and ceilings were insulated with Styrodur panels. This material reduces CO_2 emissions, as the foam cells contain nothing other than air, and contains only extruded polystyrene foam (XPS) that is free of CFSs, HCFCs and HFCs.

⏴ **0028**

Window panes and window frames were replaced with insulated versions.

0029 ➤
The remodeling of the residence improved both its appearance and the sustainable aspects of its architecture.

◄ **0030**
Large insulated expanses of glass let in more daylight while retaining heat inside.

Andrew Maynard
Architects Pty Ltd

551 Brunswick Street
Melbourne, Victoria, 3068 Australia
Tel.: +61 (3)9481 5110
www.maynardarchitects.com

0031 ➤

The approach taken for the Mash House is one that, despite first impressions, celebrates the backyard. Or perhaps less so the traditional notion of the backyard, and more so just plain, outdoor space. The original deep and dark double-fronted Victorian house offered a plethora of challenges—not least of all its lack of solar access.

0032 ▼

Our response was a series of finely crafted timber boxes nestled around the bulk of the existing house. The bedroom addition opens up the northern facade of the house to the rugged bush block, doing double duty as the roof becomes an expansive deck to extend the living space out into the treetops.

◄ 0033
More frequently, holiday homes are becoming little more than transplanted suburban ugliness; the great Australian tradition of the "shack" is in danger of being superseded by bloated mansions. With this project we wanted to celebrate the shack and have kept close to the original building's footprint to avoid taking over the rugged coastal block.

0034 ▼
Overall, the design is an experimentation of form and color, albeit a simple solution to a rather challenging suburban site. The concept was driven by obtaining passive efficiency via shrewd siting and orientation. Quality insulation, ample double-glazing and in-slab heating all combine to make this home a retro-futuristic sustainable modern house renovation.

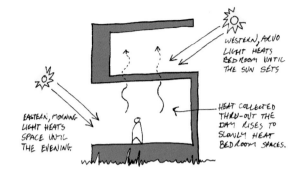

WESTERN, ARVO LIGHT HEATS BEDROOM UNTIL THE SUN SETS

EASTERN, MORNING LIGHT HEATS SPACE UNTIL THE EVENING.

HEAT COLLECTED THRU-OUT THE DAY RISES TO SLOWLY HEAT BEDROOM SPACES.

0035 ▷
The Vader House is an extension to a Victorian terrace in the dense inner city. The high boundary walls, built in disregard of existing height regulations long before such rules were created, permitted a non-standard height along the northern boundary.

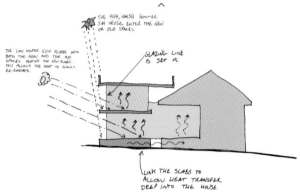

THE HIGH, HARSH SUMMER SUN NEVER ENTER THE NEW OR OLD SPACES.

THE LOW WINTER SUN FLOODS INTO BOTH THE NEW AND THE OLD SPACES. HEATING THE NEW SLABS. THIS ALLOWS THE HEAT TO SLOWLY RE-RADIATE.

GLAZING LINE IS SET IN.

LINK THE SLABS TO ALLOW HEAT TRANSFER DEEP INTO THE HOUSE.

◀ **0036**
The roofline abruptly turns to follow the dictated setback lines, resulting in a playful and telling interpretation of planning rules.

◀ **0037**
The refined palate of materials is subverted where volumes are removed to reveal the flesh inside – colored bright red with glass tiles and joinery.

0038 ➤
Recycled gray ironbark battens on the doors shade the interior from the high summer sun while allowing the lower winter rays to penetrate.

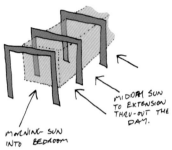

MIDDAY SUN TO EXTENSION THRU-OUT THE DAY.

MORNING SUN INTO BEDROOM

0039 ▼
Municipal requirements regarding overlooking, which dictate a 75% opacity for second-story spaces, are resolved with UV-stable stickers rather than expensive and elaborate screening.

0040 ➤
Traditional walls were replaced by bi-fold garage doors, which open the interior spaces to the long backyard, blurring the distinction between inside and outside.

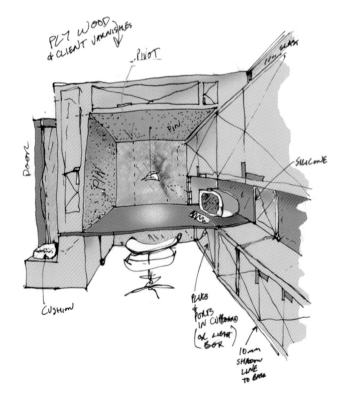

PLYWOOD & CLIENT VARNISHES
PIVOT
PIN
DOOR
SILICONE
CUSHION
PLUGS & PORTS IN CUPBOARD (OR LIGHT BOX)
10mm SHADOW LINE TO BASE

Architectenbureau
Paul de Ruiter bv

36 D Valschermkade
Amsterdam, 1059 CD, The Netherlands
Tel.: +31 20 626 32 44
www.paulderuiter.nl
twitter.com/pdrarchitects

◄ 0041
The "Kroeten island" is an artificial island with excavated waterways in an agricultural landscape. One result of this collaboration was the collective purchase of a wind turbine, which was erected 3 miles (5 km) away at the power station in Breda, The Netherlands.

0042 ►
To identify the specific wishes of all the users and integrate them into the design and to give the users a good idea of the architectural possibilities, several workshops were held during the design stage. One requirement that was specifically identified during the workshops was the need for daylight in the main auditorium.

0043 ➤

Villa Deys, Rhenen, The Netherlands
Villa Deys is a house for a senior couple. The water in the integrated swimming pool has a constant temperature of 80°F (27°C), which of course uses energy. However, by linking the pool's heating system with the low-temperature heating system of the house, in combination with a water pump, the pool area re-emits the energy it has absorbed and becomes part of the energy-saving climate-control system in the house.

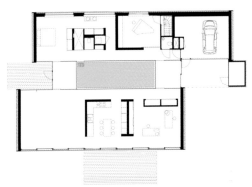

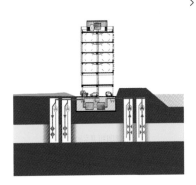

0044 ➤

Rijkswaterstaat Zeeland, Middelburg, The Netherlands
Existing techniques are used in an innovative manner to create a low-energy building, without sacrificing its economic viability or its architectural quality. The use of "active concrete" in combination with underground cold/heat storage creates a constant and comfortable working climate and results in an energy savings of 40% to 50% over traditional cooling and heating methods.

0045 ➤

All the green electricity in the TNT Centre is supplied by its "Green Machine." The bio-cogeneration plant required for this process runs on organic residual waste, among other things. Periodic power surpluses are directed to the main network, to heat the buildings in the immediate vicinity of the TNT Centre, while periodic shortages are met by taking (green) electricity from the network.

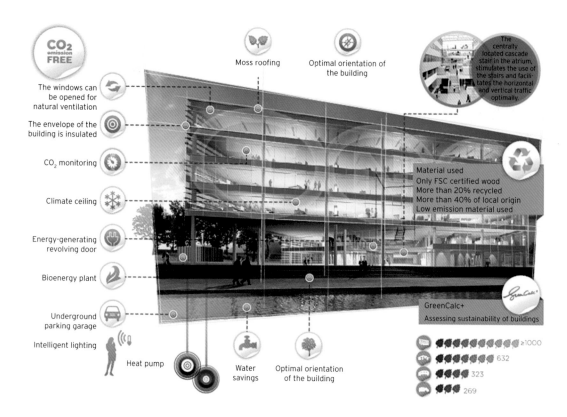

CO2 emission FREE

Moss roofing

Optimal orientation of the building

The centrally located cascade stair in the atrium, stimulates the use of the stairs and facilitates the horizontal and vertical traffic optimally.

The windows can be opened for natural ventilation

The envelope of the building is insulated

CO_2 monitoring

Climate ceiling

Energy-generating revolving door

Bioenergy plant

Underground parking garage

Intelligent lighting

Heat pump

Water savings

Optimal orientation of the building

Material used
Only FSC certified wood
More than 20% recycled
More than 40% of local origin
Low emission material used

GreenCalc+
Assessing sustainability of buildings

≥1000
632
323
269

0046 ▼

Office building, Valschermkade, The Netherlands The first sustainable decision for our new office building was to renovate an existing one. Recycled materials were used during the renovation, and the furniture was also recycled, entirely in keeping with the C2C principle. A geothermal storage system uses 20% to 30% less energy. The adjustable facade serves as a sunshade during the day and as an anti-burglary screen at night.

◄ 0047

TransPort, Schiphol, The Netherlands TransPort is the first building in the Netherlands to gain Platinum LEED certification and a BREEAM-NL - Very Good certificate with respect to its sustainability. Energy-efficient technology, such as thermal storage and concrete core activation, enhance the building's sustainability. A grass-sedum roof over the lower building provides an additional insulation layer that inhibits heat transfer.

0048 ➤

**Veranda parking garage,
Rotterdam, The Netherlands**

The primary function of the central
open area, measuring 52 x 72 feet (16 x
22 m), is to allow more light and air to
penetrate into the car park, to improve
the ambience of the underground
levels. The back pressure fans are
positioned in four exhaust tubes, which
ensures that the polluted air from the
cars is extracted from the building.

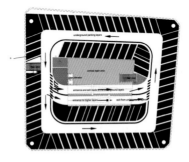

0049 ➤

**De Zuidkas, Amsterdam,
The Netherlands**

The Zuidkas is an imaginary building of
over 118,400 sq. feet (11,000 m²) with a
functional mix that is far from ordinary:
homes, offices, a school, parking
facilities, retail spaces, restaurants,
a park and a biogas electrical plant.
The objective is to make an intelligent,
self-sufficient building where energy
can be interchanged and waste flows
can be converted into heat and energy.

◄ 00050
Get inspired by sustainability.

Arkitekstudio Widjedal Racki

28 Artillerigatan
Stockholm, SE-114 51, Sweden
Tel.: +46 73 994 98 57
www.wrark.se

0051 ➤
The B-house can be joined together to form a cluster for larger camps in hostile environments.

0052 ▼
Each B-house is individualized by choosing the perfect add-ons for each location and client.

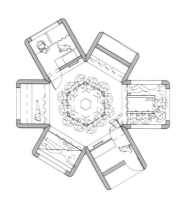

0053 ➤
Each B-house contains a floor and a roof plate. The floor plate contains most of the technical installations. These plates connect the add-ons and holds the building together. The building is a lightweight patented concrete building with waterproof surfaces and joints as part of the system. Add-ons and base plates are mounted on-site.

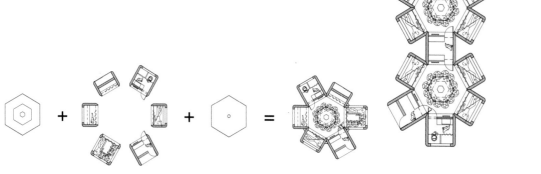

0054 ➤
The add-ons can be placed freely to maximize views and solar directions.

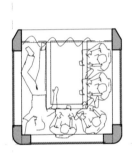

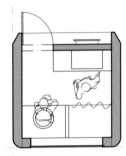

0055 ▼

The B-house's energy-positive design and recyclability addresses the need to reverse the current level of over-consumption. Materials used in its production are sourced ethically and with respect for the environment, and they are combined to create a luxurious space in which to relax and just be.

0056 ◄

An alternative use for the B-house is as a short-term solution to urban densification. Here it is being used as an easily removable complement to the 2012 London Olympic village.

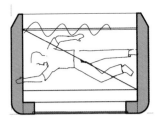

0057 ►

Individual units incorporate the highest standard of environmental design with sustainable solutions for the provision of energy, water and drainage, heating and cooling. All technical equipment is as simple as possible and is seamlessly integrated into the design. This extends to rainwater harvesting, purification and re-use, solar heating and energy, natural ventilation and refrigeration, and lighting.

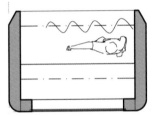

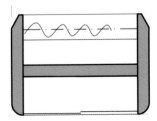

0058 ◄

Designed by internationally acclaimed Scandinavian architects, the B-House is a response to environmental and social change and provides a simple yet sophisticated solution to living and working spaces for situations where more ordinary and permanent buildings are not an option. The client's brief to the architect was to design a small, beautifully crafted eco-building that could be sited within sensitive rural environments where normal services and infrastructure are not necessarily available.

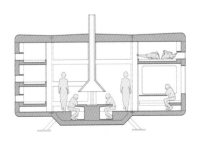

1M

0060 ▲

The building is designed to withstand hostile environmental conditions, and to provide "off-grid" living for short-to medium-term use. Its primary target market is commercial buyers and operators in the leisure and hospitality industry, as it is capable of providing eco-friendly add-on accommodation for discerning guests looking for sustainable and beautiful places to stay.

0059 ◄

A typical unit with six add-ons will have a footprint of 447 sq. feet (41.5 m²) and have one bathroom with toilet, sink, shower; one kitchen with stove and refrigerator; one storage unit with indoor as well as outdoor storage; one entrance unit with wardrobe function; one veranda room insulated or open air; 11 beds + possible additional beds in veranda room and 12 seats + possible additional seats in veranda room

Architekturbüro Reinberg
ZT GmbH

Lindengasse 39/10
A-1070 Wien, Austria
Tel.: +43 1 524 82 80
www.reinberg.net

0061 ➤

This prototype home, designed for an Austrian solar technology manufacturer, complies with requirements similar those of the PassivHaus standard, such as good façade insulation and a bioclimatic design that favors solar heat gain.

0063 ▼

The home is self-sufficient in energy. It produces all of the energy it consumes to achieve a zero-net energy balance. The south frontage gains solar energy by means of vertical openings and the picture window found on the lower level.

0062 ▲

Another active strategy is formed by the solar thermal panels integrated into the roof and the photovoltaic panels used for electricity generation.

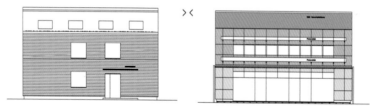

◀ 0064

In warm weather, the south-facing openings are protected from overheating by a fixed brise-soleil that forms 400 sq ft of photovoltaic panels. In winter, a heat pump provides comfort on extremely cold days.

0065 ➤

The energy for heating, hot water, ventilation, and to power home appliances and the heat pump are compensated with power from the photovoltaic panels and the heat-recovery systems, so that consumption is zero.

0070 ▼

The building is equipped with windows that open automatically only when oxygen levels are too low or there is too much moisture in the air.

◀ **0066**

Radiant heating is designed under skirting boards and flooring to heat areas of the house. The high degree of insulation achieved with triple glazing and heat-recovery ventilation reduces heat loss drastically.

0069 ▼

The house features 269 sq ft of thermal solar panels, with storage for household hot water (in the plant room and on the lower level) and a heat pump run on solar power.

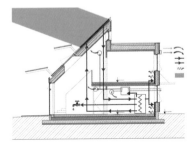

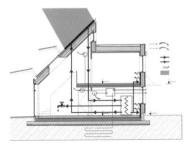

0067 ⏶

The building features a larch wood and glass skin and a roof made from a material called Eternit.

0068 ⏶

The interior walls are rendered in marl, a detritus sedimentary material with equal parts of sand, silt and clay particles.

29

Arconiko Architecten,
Frido van Nieuwamerongen /
Roswitha Abraham

Postbus 399,
Rotterdam, 3000 AJ, The Netherlands
Tel.: +31 10 4123181
www.arconiko.com

0071 ➤

Use existing knowledge of materials and encourage suppliers to change their products into C2C.
1st Cradle to Cradle House
Innovative building projects are a great chance to motivate suppliers to invest in their materials to create C2C products. All parties can eventually profit from the investment if there is sufficient publicity to show the realized effects.

0072 ▼
Use solar power for cooling

a. Geothermal heat pump to heat in winter and cool in summer. The spaces are heated and cooled by a system installed in the floor.
b. The shape of the office encourages natural ventilation throughout the building. Heat is regained by a heat-regaining unit.
c. Blow-in of cold and warm air through textile ventilation pipes with low air speed.
d. Photovoltaic cells on the south facade generate energy that can be used for the heat pump for cooling.

><

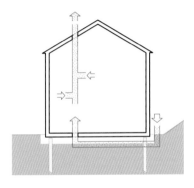

0073 ➤

More Better instead of Less Bad
Project: first Cradle to Cradle House
The design mentality of the first Cradle to Cradle house is based on a positive theory instead of fearful criticism. Since the 1970s, environmentalists have emphasized the reduction of our footprint in order to save the planet. We think that a positive mentality will be more effective and durable. Moreover, the technical possibilities have developed enough to allow us to combine comfortable living with sustainable building. The motto is to do things more better instead of less bad.

0074 ▲
Promote design over technology.
Project: first Cradle to Cradle House.
By integrating natural climatological principles within the design, high-tech installations are not needed. In the first Cradle to Cradle House, no mechanical ventilation is needed since the building itself creates a natural draft. Cold air is drained below the house through a shaft. Warm air is drafted by a chimney and rises. The wind activates a propeller that leads the air outside – it's as simple as that.

◄ 0075
Daylight as a durable design principle.
Project: Community School Rotterdam
A former academy building was transformed to house two elementary schools, two gymnasiums, a preschool and a day-care center. The inner courtyard was transformed, creating extra inside space with an atrium. The atrium brings daylight into the core of the building and serves as the central meeting place of the schools.

◄ 0076
Think positive to create a better building environment!
Project: Team Arconiko
We think in terms of opportunities rather than restrictions.

0077 ➤
Living comfort and sustainability should reinforce each other
Project: House Harsta Sky Related living and associations with a crow's nest were part of the requirements. The original house, which dates from around 1900 and was renovated in the 1960s and 1990s, can hardly be recognized after the recent rebuilding. Only the wall on the east side is reused in the modern house. The new design uses voids to create inner spaces that offer rich views that culminate in the crow's nest, a cantilevered panorama platform. Here, the purposeful daylight is crucial. The interior is designed to respect the needs of the elderly, respect the environment and maintain flexibility.

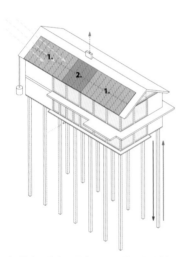

1. Photovoltaic cells for generating electricity
2. Sun collectors for generating warm water.

0078 ▲
Buildings as energy generators
Project: First Cradle to Cradle House
Buildings can function beyond self-sufficiency and even generate energy to supply the network. Photovoltaic cells on the southern inclined roofside collect sun radiation that is transformed from direct current into alternating current. This can be used directly in the house or delivered into the network for nearby buildings.

◄ 0079
Make use of existing rooftops
Project: Community school,
Rotterdam, The Netherlands

◄ 0080
Improve existing buildings
Project: Control center, Roosendaal,
The Netherlands The existing control
center of the railway building in
Roosendaal did not meet the security
demands of both its personnel and
the railway system. The technical
rooms were retained, and a new
space for the control staff was built
on top of the existing building as
a separate construction. During
the transformation, the control
center had to stay functioning and
was only stopped for 5 hours.

Arkin Tilt Architects. Ecological Planning & Design

1101 8th Street, #180
Berkeley, California, 94710, USA
Tel.: +1 510-528-9830
www.arkintilt.com

0081 ➤

The specifics of the site are the starting point. Working with the climate, landscape, views and natural grade results in buildings that are harmonious and integrated into their surroundings. On sloping sites, buildings can be literally tied to the landscape via berms and living roofs.

0082 ▽

Build as little as possible. Create multi-use spaces that can change over time. Limit or add functionality to circulation space, for example, stairways can double as libraries, landings as study or reading nooks, etc. Smaller projects use less material resources and require less energy to operate.

◀ **0083**

Employ the wisdom of previous generations. Look to local or climate-appropriate vernacular buildings (buildings built before the 1940s or without HVAC systems) for passive climate response strategies. Agricultural and manufacturing uses often yield unique and interesting ideas for storing heat or moving air.

0084 ▲
Design windows to share daylight
High clerestory windows can bring light deep into a space; salvaged interior windows can further share daylight and create spaciousness without the thermal disadvantages of single-pane glazing, older windows typically have. Films can be applied to glass in salvaged doors to achieve a required safety rating and obscure glass where desired.

0085 ▼
Buildings should primarily heat and cool themselves. Passive strategies can go a long ways toward minimizing, if not eliminating, the need for mechanical heating and cooling. In hot climates, an "ice house" roof adds a ventilated cavity above the thermal roof, providing shade for natural cooling as well as tighter overall construction.

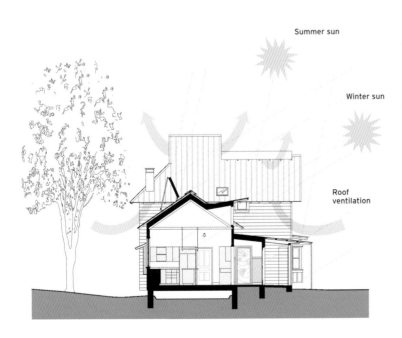

Summer sun

Winter sun

Roof
ventilation

0086 ▲
Employ natural building systems
Straw-bale construction has multiple benefits: it is an annually renewed resource that is carbon-sequestering, it has high insulation values and high thermal mass thus mediating temperature swings, it moderates humidity and absorbs sound, and it helps create a comfortable, nurturing building environment.

0087 ➤

Integrate active systems to generate hot water and electricity. Photovoltaic panels and solar hot water collectors can do double duty. For example, adjustable photovoltaic awnings can produce energy, provide seasonal shade, and facilitate the shedding of snow in the winter, all the while generating electricity. Plus, optimizing the angle increases the energy production.

0088 ➤

Maximize resource efficiency
Salvaged materials can be from a multitude of sources and are often best used in unexpected ways: Sunshades can be made from old airplane flaps, used billboard material makes a functional and playful wainscot in a school bathroom, and bowling lane boards can be reused as a table or countertop.

◄ 0089
Connectivity and delight
Introduce local materials and natural elements to reconnect to the natural world. Local soils can tie the building to its place; a tree trunk can be an elegant focal point as well as a reminder of the origins of materials and resources.

0090 ▼
Blur the distinction between indoors and out. Create outdoor living spaces or ways to convert indoor spaces to be more connected to the outdoors. Healthier and more grounded, living outdoors can re-acclimate us to the world around us. Covered porches can also provide solar control for interior spaces, west-facing ones in particular.

Atelier Riri /
Novriansyah Yakub

Jl. Perkici 12 eb 3 #19
Bintaro Sektor V Tangerang, Indonesia
Tel.: +62 812 1032303
www.atelierriri.com
atelierriri.com/blog

◄ 0091

A simple window detail in warm material such as wood can give good natural ventilation to a house. The wind can breeze through the gaps freely, doing away with the need for fans and air-conditioning.

◄ 0092

Nothing goes wrong when concrete meets vegetation. The need for safety and privacy can sometimes be resolved with a steel-patterned bar. The existence of vegetation in this equation is not merely for the sake of a soft composition. It can be a perfect base for a tree to grow on, giving a sustainable quality to the house.

0093 ►

There are many ways to provide a small house with adequate ventilation. The wall composition of cement block and glass block gives a natural and innovative solution for cross-ventilation.

◄ 0094

A water element in a house does double duty when it has a proper location and execution. The tropical climate favors its use as more than just an aesthetic piece as it can also provide a positive support to the microclimate. With the help of air movement, the surrounding semi-open spaces can benefit from a cool breeze.

0095 ➤

Industrial materials are still not popular products for use in a small urban house. In order to use such materials, the design, engineering and application must be suitable. These materials are often used in contemporary designs that respond to today's lifestyle, but the question of whether or not the material is considered sustainable may be the highest consideration.

0096 ⋀

The design for a house in a tropical climate should use the available sunlight to reduce the use of electricity during the day. A large glass wall is a commonly used design element. However, the design should avoid direct sunlight, which will raise the indoor temperature and spoil the comfort of the room. This is where a window overhang and a terrace can function as a buffer or transition.

0097 ⋀

Selecting natural materials is resourceful, so using such material is in keeping with sustainable criteria. This design attitude also narrowed the overall look into one that is suitably tropical.

0098 ➤

The use of bamboo has become more frequent in tropical-climate houses. This material is considered renewable and is abundantly available in tropical countries. With a modern frame, the room looks uniquely romantic and natural. Here a visually pleasing design is rendered in a renewable material.

0099 ⋀

Part of the house design should be aesthetical, functioning as more than just a visual element. While the sunlight nicely reflected into the room brightens the space in broad daylight, the choice of material gives a warm sensation. The parallel set of wood columns gives a play of light and shadow. while it also serves the purpose of lighting and ventilation at the same time.

◀ **0100**

A tropical climate has an abundant source of sunlight during the day in the dry season and much rain in rainy season. Therefore, the design is different from that used in other areas, such as in Europe or North America. The interior will enjoy the brightness from the sunlight, while the building design should be able to limit the intensity and duration of how the sun lights up the space. This will keep the indoor temperature comfortable.

Atelier RVL Architectes

43 Rue d'Entraigues
Tours, 37000, France
Tel. : +33 2 47 70 16 02
www.atelierrvl.com

0101 ➤
House in St.Cyr-sur-Loire (37/2007)
"Magickub" is a simple cube structure that is an "affordable, compact house, reduced to the basics, for young families."
A steel-mesh sliding door in front of the living room windows keeps out excessive heat in the summer.

◄ 0102
House in Tours (37/2009)
This addition is characterized by an extension that features the kitchen and a small dining room. The extension acts as an umbilical cord connecting the garden to the house.

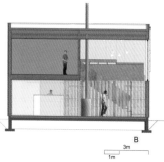

A

B

3m
1m

◄ 0103

Gymnasium in Savigné sur Lathan (37/2003)

This extension represents the remodeling of a gymnasium and locker rooms to bring them up to standards for the disabled. Rather than enlarging and making an impact on the ground, we preferred to play with the incline of the wall of the building. Made from a prefabricated wooden frame, this project of building recycling has a low impact on the environment.

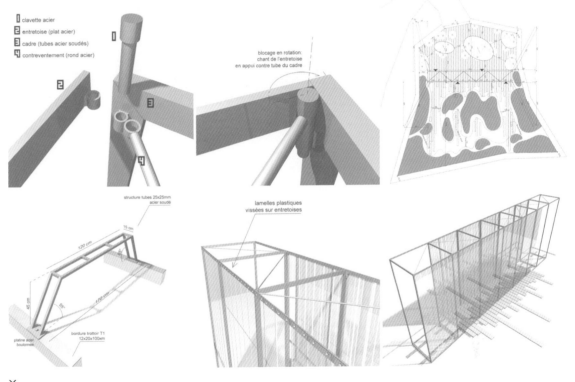

0104 ►

Jardin de Chaumont "jardin en partage" (/2008)

We explore how the notion of sharing can create multiplicity and richness, the diversity of landscapes that make up the world being a case in point. In a deliberately caricatured way, the garden also portrays the North, temperate and navel-gazing, where artifice opulence and introversion reign, and the South, which is Mediterranean and still natural, with a fecund future.

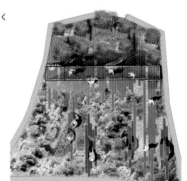

0105 ➤

Private residence meeting Passivhaus criteria, Angers, France (49/2011)
Thanks to its totally prefabricated timber frame, its double-coated insulation (cellulose wadding and wood wool), its triple glazing and its high-efficiency double-flow controlled mechanical ventilation and improved airtightness 3 cu. feet/h (0.09 m3/h), this house needs only 1 kw/sq. foot (10 kw/m²) per year.

0106 ▼

Preschool, Ambillou, France (37/2011)
A wooden lattice-type trellis sceens this preschool from the buildings opposite and provides protection from the sun. Evergreen creepers will cover the trellis. A shed-type roof allows indirect lighting in classrooms while flooding the lobby with light. It has a fully prefabricated timber frame and is covered with green roofing.

0107 ➤

Eight classrooms, one lecture hall, four youth clubs and 30 rooms for boarders of Loches (31/2011)
Over-insulation added to a wood concrete hybrid frame coupled with double-flow ventilation reduced construction time and the impact of the building on the environment.

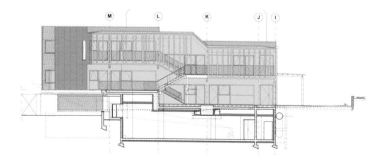

0108 ➤

This contemporary house is a hybrid building that fits perfectly into the future suburban fabric. It takes advantage of the steep ground while looking over the magnificient views of the site. This house is based on a design for a barn and was adapted to be suitable for a home on a rather unusual site. The house is actually a series of four treillis-like gables.

><

◄ **0109**
House in Tours, France (37/2009)
Covered in wood and functioning as a green-house, this section of the building allows the existing house to take in a maximum amount of sunshine. The wooden solar protection can be removed during winter and put back during the summer.

◄ **0110**
House in Fondettes, France (37/2011)
Surrounded by mature trees, this construction, built on a northern hillside has large windows on the south side, taking bioclimatic principles into consideration.

Baumraum / Andreas Wenning

49 Roonstrasse
Bremen, 28203 Germany
Tel.: +49 421 70 51 22
www.baumraum.de

0111 ▼

Tree houses are fragile structures supported by living things, which means that the useful life of the house depends to a large extent on the strength and life of the tree it is built in.

0112 ▲

From inside the cabin, thanks to wall-spanning windows and a skylight, the occupants can freely enjoy nature, which is literally within their grasp.

0113 ⬆

The dream of living in and with nature comes true in this type of home, which puts most of its weight on the resilient oak trunk.

0114 ▼

Oak is the predominant building material. The railings that limit the habitable space are made of oak and steel.

0115 ➤
The side walls are covered in oak panels over rockwool, which serves as both insulation and wind foil.

◄ **0116**
The name of the project—Between Alder and Oak—simply responds to the fact that the tree house is built between two different species of tree: an alder and an oak.

◄ 0117
The oak that partially supports the load was not strong enough, so the architect placed two support posts over a concrete base for security.

0118 ▲
At a height of 13 feet (4 m), there is a platform-terrace with a table and a few chairs and an upper cabin with a curved roof clad in wood and bitumen.

0119 ▼
A few steps lead up from the terrace to the cabin.

0020 ►
Built for a family with grown-up children, this wooden cabin—16 feet (5 m) above the ground—provides its residents with space for rest and relaxation, and the possibility of adapting one of the compartments for use as a guest room.

Beals - Lyon Architects

119 Sinclair Road
London, W14 0NP, UK
Tel.: +44 20 7610 4638

Monseñor Felix Cabrera
23, of. 33, Providencia
Santiago, Chile
Tel.: +56 2 2318824
www.beals-lyon.cl

0121 ▼

From objects to ecologies

In times when architectural production
is governed and judged by issues of
external appearance and originality
of shape, we would like to think
of sustainability as a way to move
our understanding of architecture
from the production of isolated and
autonomous objects, to one focused on
relationships to context. A new ecology
that emerges from more diffuse,
vague, or ambiguous boundaries.

0122 ▲

Blend-in with surroundings

For a sustainable design, it is
important to understand and know
the place and be responsive to local
conditions. By borrowing shapes and
textures of existing buildings in the
area, a new construction can blend
in with its surroundings, creating
a fragile image and appearing as
if it has always been there.

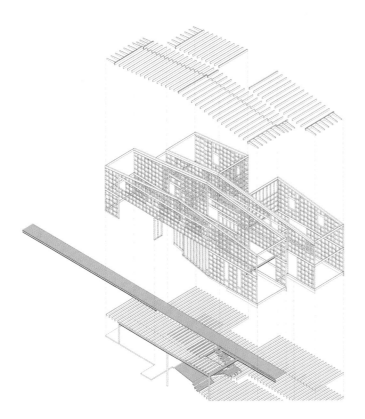

◄ 0123
Use local material and techniques
When working in remote and isolated sites, using available materials and building methods significantly reduces the environmental impact of new constructions. If using renewable resources, timber is the most sustainable alternative.

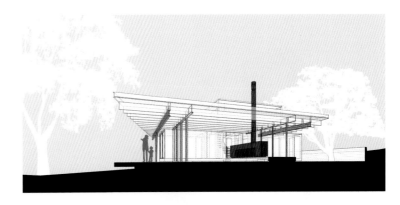

0124 ▲
Create an in-between space
Not an interior, nor an exterior, but something inbetween. An in between space blurs the boundary between interior and exterior, creating a more ambiguous relationship between building and landscape, natural and man-made.
© Courtesy of Beals – Lyon Architects

◄ 0125
Do not design objects, but environmental conditions
A roof that extends beyond the interior provides a surrounding shadow, creating a more temperate environment.

0126 ➤

Build as light as possible.
The fewer materials employed,
the more sustainable the building.
Thus, structure is the result of the
search for efficiency in performance,
which is frequently found in older
and traditional constructions.

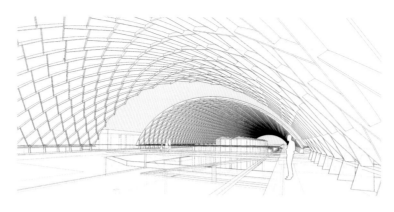

0127 ▼

Touch the ground lightly
Projects must emphasize nature by
respecting the landscape, creating a
new approach to it, not against it.

0128 ➤
Erase architecture by reflection
By means of reflection, it is possible to disappear in the landscape. The Puddle Fountain resembles a puddle in a park, mirroring the sky, creating a looking glass on the damp ground.

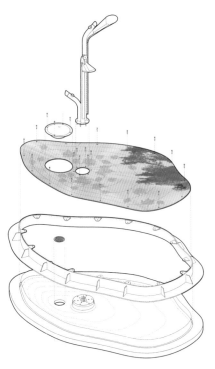

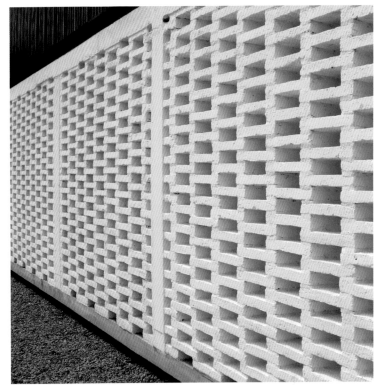

0129 ▲
Use porous boundaries
More permeable walls allow the exterior to migrate into the interior and vice versa.

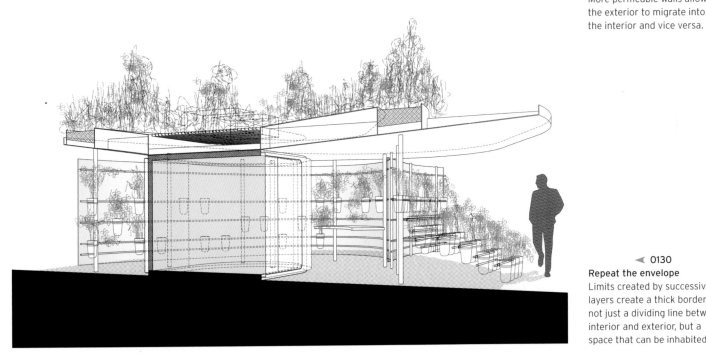

◄ 0130
Repeat the envelope
Limits created by successive layers create a thick border. It's not just a dividing line between interior and exterior, but a space that can be inhabited.

Bipolaire Arquitectos /
Miguel Arraiz García

Tel.: +34 963 476 566
www.bipolaire.net

0131 ➤

In situ design
We decided, on a trial basis, to make
partially open-ended decisions,
so as not to have to define
everything in the project phase.

0132 ▼

Accessibility
Accessing the home, located on a level
below the entrance level of the lot,
should always be possible, even in the
long-term, anticipating the possible
decreased mobility of the residents.

0133 ▲

Materials
Suitable materials were chosen
depending on their location,
how they will be used and their
manufacturing process.

◄ **0134**

Energy
Use geothermal energy for heating,
solar energy for hot water and gas
for cooking. The combination of good
insulation in the facade and roof,
properly thought-out air circulation
and the use of renewable energy are
obligatory for a contemporary home.

0135 ▽
Water
Rainwater is collected and stored. Gray and sewage water is purified and stored. Rainwater and purified water are mixed and reused for watering. Water becomes an important element of the garden, producing a touch of freshness, sounds, reflections, etc.

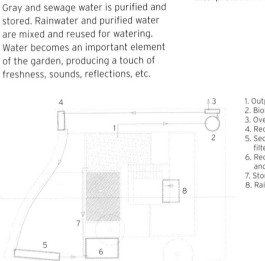

1. Output for gray and black water
2. Biological treatment
3. Overflow to sewers
4. Recycled water cistern
5. Second phase of treatment: sand filters and vegetation (reeds)
6. Recycled water tank and storm-water
7. Storm-water outlet
8. Rainwater cistern

0136 ➤
Tree-building relationship
Trees are a starting point for a project. Each opening in the facade is linked to an existing or new tree that frames views, protects from excessive sun and provides the end perspective and interpretation of the environment.

0137 ▲
Roof garden
It acts as the threshold of the landscape, banishing the nearby views of roofs and embracing distant views. It re-states the green surface occupied by the plan of the home and provides excellent insulation.

0139 ▲
"Time is money" is not compatible with sustainable development.

0140 ➤
Reusing waste
A slow working pace can reduce the amount of waste generated to an unimaginable minimum. The few remnants of the materials used are carefully stored and reused in subsequent stages of the construction. The 4,090-sq.-foot (380 m²) project generated 388 cu. feet (11 m³) of waste.

0138 ➤
Natural resources
Cross ventilation, exposure to prevailing summer winds and protection from prevailing winter winds provide a comfortable indoor climate with a very stable temperature.

BNIM

106 West 14th Street, Suite 200
Kansas City, Missouri, 64105, USA
Tel.: +1 816 783 1582
www.bnim.com

0141 ➤
Add vitality
Every act of building must add vitality and resilience to the surrounding community and ecosystem. Sustainability does not have to be exclusive of design excellence. Each must coexist for either to be achieved. In Greensburg, Kansas, after the entire town was destroyed by a tornado, they vowed to build back better and stronger.

0142 ▲
Promote obsolescence
We can design buildings, landscapes and communities that are compelling enough to make our current ideas of community obsolete. These new designs improve the health and quality of life of each inhabitant, while increasing the vitality of our environment.

0143 ▲
Unite human with nature
Buildings are for people. Sustainable
architecture should celebrate
beauty and connect us to our innate
knowledge of the living world.

◄ **0144**
Reuse existing building stock
The most sustainable construction
projects reclaim from the vast
inventory of existing stock to
create new life and purpose.

0145 ➤
Generate more than you consume
A building can give back more resources than it consumes through carefully orchestrated strategies designed to achieve a net zero impact. In a cycle of restorative design, a building can renewably generate energy to operate (and often with a surplus); capture, treat and use its own water (and that of surrounding buildings); and operate using only what the site can provide.

 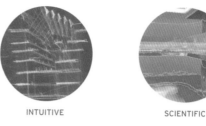 

INTUITIVE   SCIENTIFIC   EXPERIENTIAL

sketch   research   investigation   modeling   construction

0146 ⋀
Use intuition and science
Though we begin with intuitive design ideas, scientific engineering and modeling help us predict how that intuitive approach will perform in a given climate. We often retool the design to make the most of the building technologies.

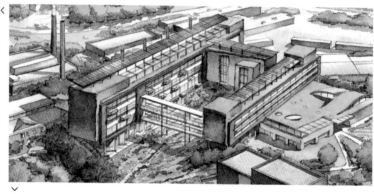

◄ 0147
Building as pedagogy
Look at buildings and sites as teaching tools. Each element of a project has the pedagogical potential, filled with the promise of imparting knowledge to those who interact with it.

◄ **0148**

No one knows as much as everyone
Collaborative intelligence creates a sense of ownership, facilitates the sharing of resources and allows the best ideas to emerge. Whether it's collaboration within an integrated design team or a community vision that emerges from social media tools, using a collaborative dialogue of discovery can and does produce the best ideas.

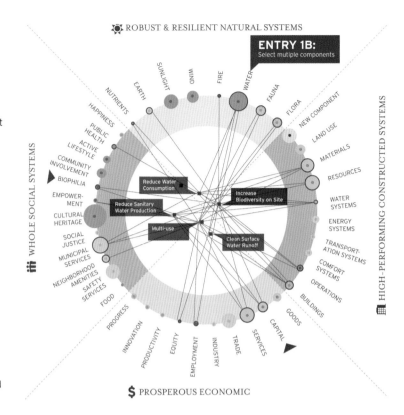

ROBUST & RESILIENT NATURAL SYSTEMS

ENTRY 1B:
Select mutiple components

WHOLE SOCIAL SYSTEMS

HIGH-PERFORMING CONSTRUCTED SYSTEMS

Reduce Water Consumption

Increase Biodiversity on Site

Reduce Sanitary Water Production

Multi-use

Clean Surface Water Runoff

$ PROSPEROUS ECONOMIC

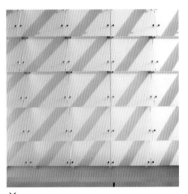

◄ **0149**

Analyze and measure
High-performance design for new buildings–and sustainable improvements to existing buildings–achieve quick returns through reduced utility costs, increased productivity and improved social and corporate equity.

0150 ▲

Plus ultra
Latin for "more beyond," this mantra encourages us to challenge what we know, go deeper to ask questions and find answers and ultimately, to seek ways of building that are restorative and symbiotic with nature. The Packard Foundation Sustainability Matrix and Report and the REGEN Tool from the U.S. Green Building Council are two tools that help designers make more informed decisions.

Bourne Blue Architecture

PO Box 295
Newcastle, NSW, 2300, Australia
Tel.: +61 02 4929 1450
www.bourneblue.com.au

◄ 0151
Design buildings well, so that they are well used and do not become obsolete as the client's needs change. Good design will be timeless, and delightful spaces will always be fully utilized.

0152 ▲
If possible, recycle buildings rather than knocking down existing structures and rebuilding. Look for the positive elements within the existing buildings and reinforce them. This building was a small seaside cottage that was pulled apart and converted to a living space with new sleeping spaces added as pavilions in the yard. Where walls were removed, contrasting floorboards were added so the previous entity could always be understood.

0153 ▼
Design a building to use minimal energy. Incorporate solar panels. Use low-energy light fixtures, hot water heaters and appliances. In a temperate climate, orient the building east-west to maximize the ability to catch summer sun.

0154 ▲
Position openings so they will allow the occupants to maximize cross-ventilation in summer. By placing large louvered panels close to the ceiling at both ends of this hall, convective currents will always force air out on hot days.

0155 ▽
Collect all rainwater and use on-site to flush toilets and water gardens.

0156 ▲
In temperate climates, insulate the building well to reduce heat loss in winter and heat gain in summer. This project has insulation (both bulk and reflective) in all wall and ceiling cavities.

0158 ▲
In temperate climates, incorporate floor slabs that are shaded in summer and flooded with sun in winter. This way the floor stays cool in summer, but as more sun hits the surface later throughout the year as seasons change, it radiates more heat into the room.

0157 ▽
Use materials that have minimal environmental impact in their manufacture, such as reconstituted or plantation timbers, recycled polyester insulation, plywood and low-VOC paints.

0159 ▽
Encourage outdoor living through a partially covered "in-between" zone. Spaces not fully inside, nor outside but protected require little energy to sustain and provide a delightful living space.

0160 ▷
Install screens or shutters, and large opening doors and windows so that the building can be "trimmed" like a sail on a yacht. These shutters can fully close, fully open or sit somewhere in between, allowing the inhabitants to adjust the building to suit the weather.

Casagrande Laboratory Taiwan

Department of Architecture,
Tamkang University
151 Ying-chuan Road
Tamsui, Taipei, Taiwan
Tel.: +886 2 2621 5656
www.clab.fi

◄ 0161
Can't push a river
Chen House is raised above the ground in order to let the frequent heavy storm waters run free under the house. The house is not an obstacle in the natural flows of energy. (Chen House, Casagrande Laboratory, Taiwan, 2008)

0162 ►
Break the box
Modern man in an air-conditioned box is a degenerated animal. Bio-climatic architecture aims to break open the modern box in order to let nature in. Chen House's gaped walls are essential to the bio-climatic architecture. (Chen House, Casagrande Laboratory, Taiwan, 2008)

Local knowledge
Local knowledge builds a connection between the modern human and nature. Industrial cities are alien to the local knowledge, which pushes the built human environment toward the organic. (Bug Dome, WEAK! [Marco Casagrande, Hsieh Ying-Chun, Roan Ching-Yueh], Shenzhen, 2009)

0164 ➤
Insect Architecture
To design is not enough. Design should not replace reality. The building must grow from the site; it must react to its surroundings, it must reflect life and it must be left to be itself, like any other living creature. Architectural control is against nature and it is against architecture. (Bug Dome, WEAK! [Marco Casagrande, Hsieh Ying-Chun, Roan Ching-Yueh], Shenzhen, 2009)

0165 ⬆
Dominate the no-man's-land
This same space is shared by humans and nature, human nature as part of nature. In the organic settlement of Treasure Hill, the balance between humans and jungle is constantly changing. This is a living ruin. (Treasure Hill, Casagrande Laboratory, Taiwan, 2003)

0166 ➤

Existence maximum

The only rule of nature is to produce maximal life. House and city should be part of the life-providing system, not an enclave in it. The industrial city has many rules. (Ruin Academy, Casagrande Laboratory, Taiwan, 2010)

0167 ▼

Ultra-ruin

Ruin is when man-made has become part of nature. Modern architecture and the industrial city must be ruined. You can drill holes through the building in order to let rain inside. (Ruin Academy, Casagrande Laboratory, Taiwan, 2010)

0168 ➤

Caveman style

Modern humans have to take the liberty to travel 1,000 years back in order to realize that things are the same. (Ruin Academy, Casagrande Laboratory, Taiwan, 2010)

0169 ➤
Flesh is more
Architecture is a meditative space between modern humans and nature. Site-specific knowledge is gained through physical labor. (Floating Sauna, Casagrande & Rintala, Norway, 2002)

0170 ▾
Land(e)scape
You can build legs for abandoned farm houses so that they can follow the farmers to the cities in the south. (Land(e)scape, Casagrande & Rintala, Finland, 1999)

Center for Design Research, School of Architecture, Design & Planning, University of Kansas

Lawrence, Kansas, 6645, USA
Tel.: +1 785 864 2700
www.saud.ku.edu

0171 ➤

This program is run by the University of Kansas for graduate students in its Architecture and Urban Planning program. It focuses on prefabricated building systems and responds to the general problems of adapting building prototypes to suit local contexts.

0172 ➤

Located in Kansas City's Rosedale neighborhood, this home is ideal for those who want to live "off the grid" without missing out on the amenities offered by living near the city's urban core.

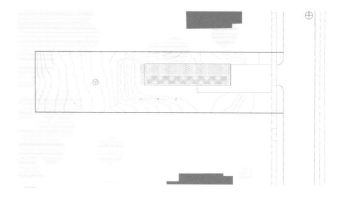

◄ 0173
The program aimed to achieve LEED Platinum certification for the house and to be the first home in Wyandotte County, Kansas, to use renewable energy.

0174 ▲
The house is designed to produce surplus energy and sell it back to the grid. The roof features 600 sq. feet (56 m²) of photovoltaic panels. These are complemented by a wind turbine and a geothermal energy source. The cellulose insulation in the walls and roof has a value of R-20.

◄ **0175**
The solid hardwood timber used on the exterior walls was sourced from South America and is FSC certified. Interiors and the frame make use of 70-year-old recycled Douglas fir.

◄ **0176**
The Sustainable Prototype was built and delivered a year after a tornado devastated the Greensburg area. The design responded to the specific brief of the 5.4.7 Arts Center, the principal client, and was intended to act as a community center for the area.

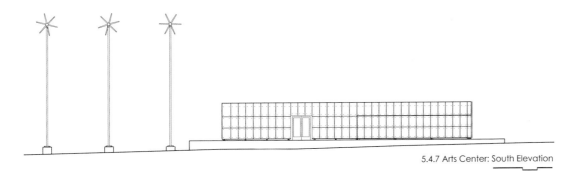

◄ 0177
It features wind turbines that supply an estimated 430 kWh/month of energy, for a mean wind speed of 11 miles per hour (18 km/h). Photovoltaic solar panels contribute 1.4 kWh.

5.4.7 Arts Center: South Elevation

20 10 5 0

0178 ▼
Greensburg is located in Kiowa County, in southwestern Kansas, and has a large farming sector. In May 2007, the town was devastated by a tornado that destroyed 95% of its buildings and left 11 people dead.

0179 ▲
According to the student team that designed it, the chief environmental positives of the prototype were its seamless blending with the landscape, its strategic use of active and passive strategies to meet energy requirements (80% to 120% depending on the season and wind speed), its use of recycled materials, and its flexibility for use as an exhibition space, community center or space for offices.

0180 ▼
Although the Arts Center did not require public funding, they chose to follow the guidelines for publicly funded buildings as an act of responsibility and became the first building in Kansas to receive LEED Platinum certification.

Christina Zerva Architects

Ossis, 3
Larissa, Greece
Tel.: +30 2410 258003
www.christinazerva.com

0181 ▼
An operable roof system allows the cool air to enter and the warm air to escape through the opening, thus perfectly ventilating the space. You can watch the stars at night or sit snug in a rain shower. It adds value and enhances the living space or, in this case, the dining area of the restaurant. When open, the natural light illuminates the dining area, creating a special atmosphere, while also providing energy savings.

0182 ▲
The facade is equipped with smart glass and a movable shading system in order to comply with thermal comfort requirements. Creating a "moving air chamber" between the wood panels and the smart glass prevents condensation from forming behind the panels and provides thermal and acoustic insulation. The airflow through the glazing cavity is driven by natural buoyancy and aided by wind pressure differences.

0183 ➤

Green roofs are biological sponges and filters that regulate building temperatures by acting as insulators, increase urban wildlife habitat and reduce the urban heat-island effect in our cities by dramatically absorbing and slowing the rate of storm-water runoff from buildings.

0184 ▽

Environmental design considerations shape the architecture of advanced naturally ventilated buildings. By splitting the building in to two separate cubic forms and lifting it up from the ground level, the force of the wind splits too. This contributes to cooling of the floors and roofs of the building.

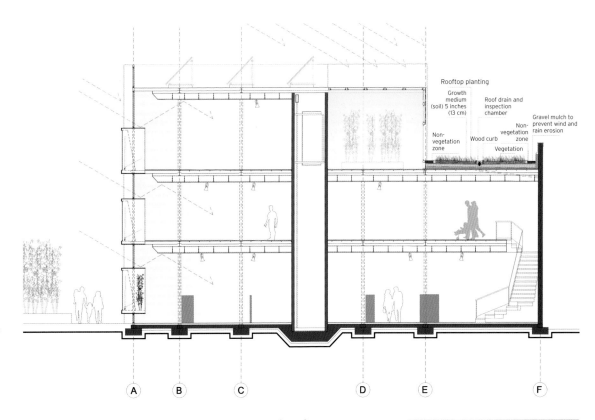

Rooftop planting

Growth medium (soil) 5 inches (13 cm)

Roof drain and inspection chamber

Non-vegetation zone

Wood curb

Non-vegetation zone

Gravel mulch to prevent wind and rain erosion

Vegetation

A   B   C   D   E   F

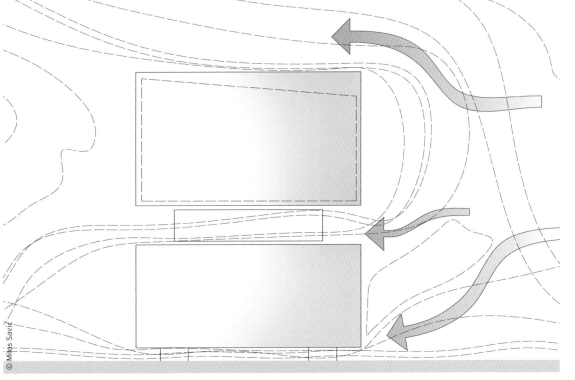

© Milos Savić

0185 ▼

Eco-friendly building materials provide environmental and health benefits that traditional materials typically don't. The production and application of these materials means less energy consumption and less natural resource depletion and pollution. They are also generally less toxic for the planet and its inhabitants. Green materials are: reusable, renewable, recycled, durable, locally available and low- or non-toxic.

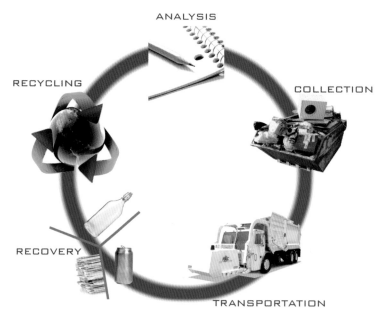

ANALYSIS

RECYCLING

COLLECTION

RECOVERY

TRANSPORTATION

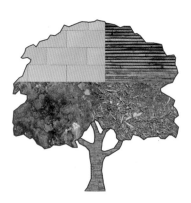

0186 ◄

Design decisions can make a significant contribution to preventing and reducing waste in the first place. Purchasing materials with less or returnable packaging and pre-ordering materials to specification reduces waste. It is very important to reuse any excavated soil to landscape other areas. Applying this philosophy early on can dramatically change and ease the process of waste management.

0187 ►

The role of shades is very important and is one of key passive design elements in eco-friendly architecture. These wooden shades can be spaced to maximize winter sun penetration, allowing passive solar heating, and still allow protection in summer to minimize heat gains. Adjustable shading is recommended for elevations that receive a combination of high- and low-angle sun throughout the day and year.

0188 ➤

Vegetation should play a significant role in architectural design, due to the many advantages it offers to the internal environment, the occupants and overall aesthetics. Vegetation systems become an active cooling agent for buildings, reducing the need for air-conditioning. Indoor plants purify, humidify and oxygenate air, improving indoor air quality and also adding a psychological value.

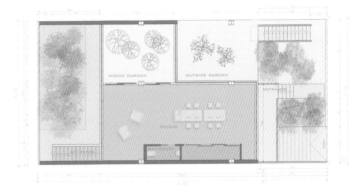

0189 ▼

Solar-performance glass can help prevent heat gain and insulate against heat loss, improve thermal comfort and reduce glare and UV radiation. In the winter months, much of the warmth given off by the heaters inside the building is bounced back into the room. In the summer, the sunrays bounce off the glass and the heat stays outside. Eco-glass creates a healthy and comfortable environment.

0190 ▼

Window performance and orientation are crucial for determining the amount of solar radiation a house receives. The main south-facing facade, with the large angled opening, receives more solar radiation than a vertical one would thus capturing as much solar energy—free heat—as possible, for best energy saving winter performance.

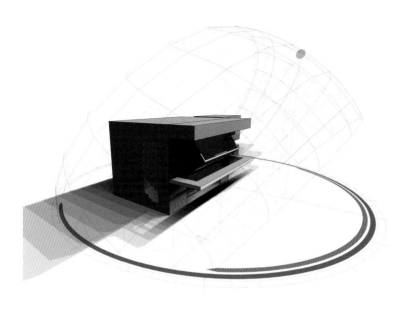

Correia / Ragazzi Arquitectos

Graça Correia, Lda
39 Rua Azevedo Coutinho, 4º Sala 44
Porto, 4100-100, Portugal
Tel.: +351 226 067 047
www.correiaragazzi.com

0191 ➤

The house opens itself in all directions in order to maximize the views of the beautiful vineyards as well as the amount of sun exposure (with adequate shading on the sunny side, preventing the need to use cooling systems).

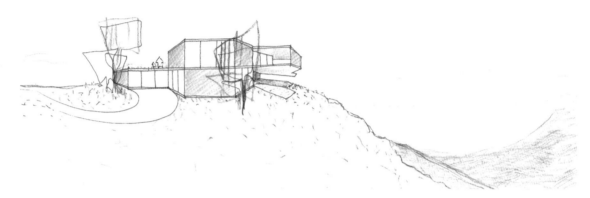

◄ **0192**

The project's positioning on the lot was essential given its surroundings. The house automatically dominates the landscape, but this dominance should also be an element of the house, which frames the landscape, as in a painting.

◄ 0193

0193
No trees were cut down to build this house, in a national park. The house overhangs a riverbank cliff, reducing the building's footprint on the land. From the river it appears as a clear glass frame that blends in with the vegetation, making it almost imperceptible.

◄ 0194
Two tiles that reinvent the Portuguese tile tradition. This one is produced from recycled materials and leftovers of tile production. (Shown in high and low relief.)

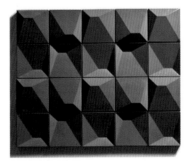
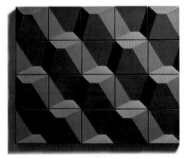

0195 ►
From both tiles, either together or separately, one can create several different surfaces of high plasticity. Countless patterns and textures can result from different combinations.

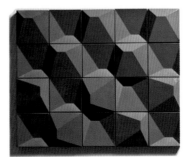
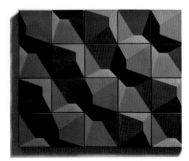

The courtyard establishes urban relations and allows the transition between outside and inside private housing spaces, creating a new layer of privacy and occupancy. The platforms camouflage the lot's natural undulations, allowing the artificial landscape to complement the natural one, composing a new landscape idea. Vegetation and water are introduced as means of controlling the heat-island effect, providing shading and cooling and diminishing pollution.

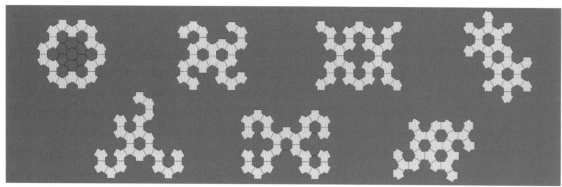

◄ 0197
A system of cell growth
Each individual cell is an apartment that can be used in a honeycomb composition that can be increased according to the housing needs. Flexible layouts allow adjustments to accommodate the topography, sun exposure, ventilation, areas, etc.

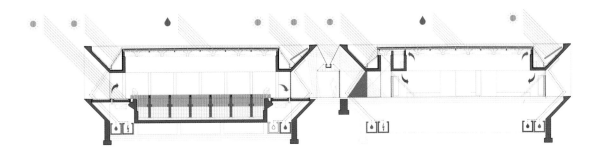

◄ 0198
In a big public structure, it is of vital importance to be as sustainable as possible in term of energy efficiency and the use of natural resources. Also, the use of prefabricated elements improves the construction's cost efficiency both in terms of budget and also in time spent.

0199 ▼
SYN comes from the Greek and means "with" or "together." SYN is both a source of light and ventilation, one piece that combines both functions and is ideal to use when renovating old buildings and in small rooms, such as bathrooms. SYN is made from 90% recycled material.

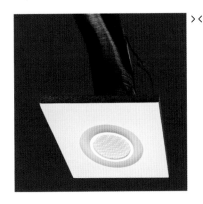

0200 ➤
The existing ruin was rebuilt and is now a guest bedroom; this solution allowed for another bedroom so the entire family could be accommodated without needing to increase the waterproof area in the lot, in an area of heavy rain.

Coste Architectures

11, Rue de la Prévoté, BP 19
Houdan, 78550 France
Tel.: +3 1 30 59 54 95
www.coste.fr

0201 ▼
This property incorporates thermal solar panels, located on the roof, which provide over 50% of the energy required to heat the water used in the home.

0202 ▲
This latest model for a prefabricated home was designed by Agence Coste Architectures, but the actual construction is the responsibility of the Geoxia consortium (Maisons Phenix).

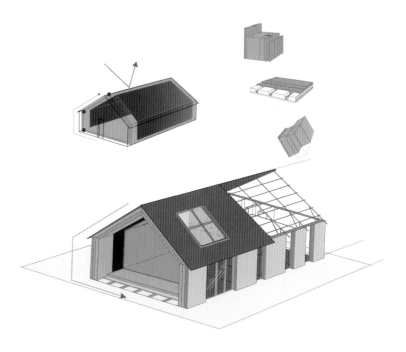

◄ **0203**
From the outside, the residence appears to be made up of fairly classic cube forms: a rectangular shape and gable roof. The outer cladding on the walls is wood.

◄ 0204

Thanks to the design of its layout and all its other bioclimatic features, the property does not need to be heated by electricity. To achieve optimum insulation and prevent heat loss, the project integrated thermal bridges and triple-glazed windows.

◄ 0205

In addition, the walls have a double R-value in terms of thermal resistance, in accordance with French technical regulations.

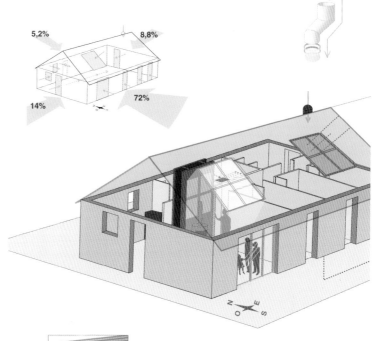

0206 ►

From the perspective of sustainable architecture, the house offers a multitude of solutions for greener living. The first is its optimum orientation to the sun, since 72% of the natural light penetrates the building on its southern side.

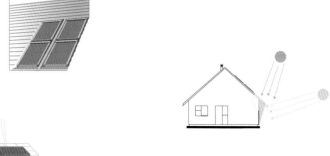

0207 ➤

The energy consumption for the entire house is about 1 kW/sq. foot (12 kW/m²) per year, whereas in normal conditions, without taking any measures to increase energy efficiency, it would amount to roughly 6.5 kW/sq. foot (70 kW/m²) per year.

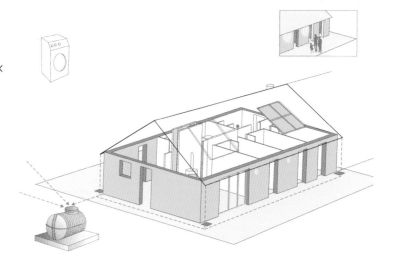

➤ ➤

◄ **0208**

The air passes through a double-flow thermodynamic system and is distributed by two diffusers that help to renew the atmosphere inside the house throughout most of the year. A wood-burning stove has also been installed to enhance the heating system.

◄ 0209
An earth tube (an earth-air heat exchanger) buried in a trench at a depth of 8 feet (2.5 m) feeds a ventilation system with air at a constant temperature of 57°F (14°C) (with variation of +/-3.5°F or +/-2°C).

0210 ►
These energy savings, which are calculated to be around 80%, and for the water estimated at about 38%, translate to a reduction in CO_2 emissions into the atmosphere of some 2,755 pounds (1,250 kg) per year – figures that are by no means negligible, showing that reducing emissions is a viable position.

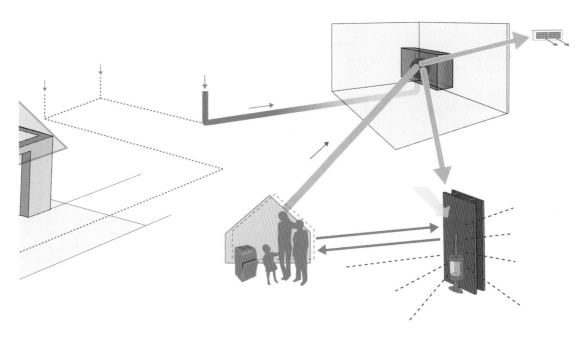

Designs Northwest Architects

10031 SR 532, Suite B
Stanwood, Washington, 98292, USA
Tel.: +1 360 629 3441
www.designsnw.com

0211 ➤
Limit the footprint
Building small reduces the impact on
the existing ecosystem by allowing a
larger amount of natural vegetation
to remain in tact. A smaller house
also uses less energy and leaves
a smaller carbon footprint. It
also uses less material, saving
resources, which is beneficial to
the environment as a whole.

0212 ▼
Use passive cooling
You can reduce the need for
mechanical cooling by using passive
strategies. Windows can be located to
provide cross-ventilation. Sun shading
devices can be used to reduce solar
gain. Operable walls can be used
to create semi-outdoor rooms.

◄ 0213
Use concrete floor systems
Concrete floors can be used as a
thermal mass to store energy gain
during the day and release it during
the night. This moderates drastic
temperature swings and reduces
the need for mechanical heating
and cooling. Concrete floors can
also be used for hydronic heating
which can be as much as 30% more
efficient than a forced-air system.

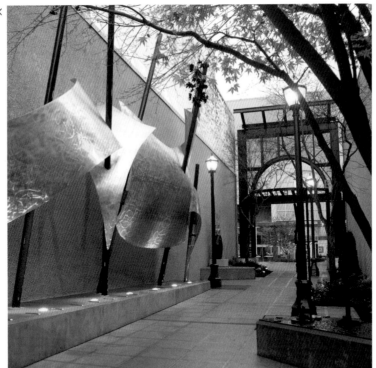

◄ 0214
Borrow light for dark spaces
Take advantage of natural light whenever possible. It saves energy and is also beneficial for psychological health. By using skylights, large windows and delicate structures, you can bring light into interior spaces. Bright colors or reflective materials can also be used to reflect natural light farther into darkened spaces.

00215 ▼
Use native landscaping
Preserve as much of the native vegetation as possible and look for opportunities to restore areas to their natural state. Climate-appropriate plants will need less irrigation and be easier to maintain. They will also be beneficial to the surrounding ecosystem.

00216 ▼
Prioritize your use of glazing
Orient major glass openings to views or to a desired sun exposure and limit the amount of glazing in other areas. Prioritizing in this manner will allow you to reduce the overall amount of windows, making the building envelope more efficient.

0217 ➤

Use a rain screen system

A rain-screen cladding system creates a breathable air space between the siding material and the air barrier/drainage plane. This prevents moisture from being trapped behind the siding, which increases the longevity of the materials and helps prevent the development of toxic mold and mildew.

0218 ➤

Choose materials carefully

Sustainable materials, such as engineered wood products, composite materials and steel, often use less raw material and are more durable than traditional building materials. Also look for opportunities to salvage and reuse materials. Try to think beyond just the aesthetic when choosing your materials.

◄ 0219

Adaptive reuse

Existing buildings can often be salvaged and repurposed for new uses. Even if the entire floor plan, aesthetic character, and use of the building are being changed, saving elements such as the foundation or exterior walls can eliminate a lot of disruptive site work and keep materials from reaching landfills.

> <

0220 ➤

Green roofs

Green roofs have many sustainable properties. They provide superior insulation without using synthetic material, they retain and filter storm water runoff, and they provide a micro-habitat for various native species of flora and fauna.

Despang Architekten

1040 North Olive Road
PO Box 210075
Tucson, Arizona 85721, USA
Tel.: +1 520 626 9350
www.despangarchitekten.com

0221 ➤

Ecology as a universal system requires a comprehensive architectural methodology that facilitates human events that inform spaces, shaping their form. The individual ecological human activities create typological prototypes that, together, form a holistic bioclimatic way of living. A fundamental type is movement between public transportation spaces, riding in style with architectural aesthetics engaging the senses.

◄ **0222**

This grocery store is decentralized but still at a walkable distance, which encourages customers to have a healthy attitude. The Hannover Jibi grocery store has a green vegetative facade, which supplies the majority of the store's energy by way of its PV roof, and a beverage pavilion that has a textile cloud hovering over it, providing shade and creating a biodynamic spectacle.

0223 ▲

To get people out of their anti-social and counter-environmental personal transportation, public transit must be sexier. The subway canopies for the city of Bochum, Germany, adjust the eye from dark to bright and provide a low-maintenance texture. The Karlsruhe signposts deliver the information to the traveler with an uncluttered minimalist approach.

0224 ➤

The next three tips represent dwelling types in relation to the most ecological building being one that is not newly built but rejuvenated and kept in the life cycle. Here, climate change serves as inspiration for innovating traditional rooftop sleeping in the cool summer breezes. The Hannover case study interprets Schindler´s theme as an assemblage of thermally modified timber.

0225 ▲

Considering that 90% of the buildings believed to be standing in 2030 are already in existence, the Hannover project's ecological mission of rejuvenation advocates thermal improvements to the building envelope as a means for responsible remodeling. An unfinished attic is morphed into a loft and composes outdoor spaces around a typical 1960s multi-dwelling house.

0226 ➤
The most effective means of infusing ecology into society is through educational institutions. This allows a new building to emerge in an environment that is open to architectural and environmental concerns while simultaneously having them disseminate the environmental message. The Ilsede project adapts an existing school for all-day use by offering a universal space within a bio-regulated exoskeleton.

0227 ▼
The Göttingen University Passive House kindergarten is a monolithic concrete structure that is a thermal-massing hybrid of landscape and architecture. It radiates out toward the south, fabric shades offering unobstructed views from the windows.

◄ **0228**
The first Passive House kindergarten in Hanover, Germany innovates the U.S. frame system by deepening it with TJI wall trusses and cellulose insulation. The curvilinear glazed south facade optimizes the group's intimacy, seamlessly transitions into the outdoor classrooms and maximizes solar gain. The other three facades soften their hermetic thermal enclosure with slices of light.

0229 ▲
Reviving the ecological relationship of working close to home in both urban and rural settings, the "Headquarters Krogmann" exemplifies transitioning from a traditional small-town wood contractor into the 21st century. In plan and section, the cone-shaped structure demonstratively embraces the sun and the center of the small town, heralding a Passive House/ Active Working environment.

◄ **0230**
A 300-year-old farmhouse gets its embodied energy expanded through a critical restoration. Strategic parts of the half-timbered loam brick infill were replaced with a new spatial and thermal threshold of triple-paned glass. Recessed from the exoskeleton, the glazing creates transitional outdoor spaces and, internally, features a hybrid of Miesian "Free Plan" and Loosian "Raumplan."

Dick van Gameren Architecten

4B Willem Fenengastraat
Amsterdam, 1096 BN, The Netherlands
Tel.: +31 (0)20 4627800
www.dickvangameren.com

0231 ➤

Reuse
The design steps off from the existing house so that, along with comprehensively improving the quality of both space and building performance, it makes the most of the materials already on-site. Components of the existing house that had to be removed have, where possible, been reused elsewhere in the design.

0232 ▼

Energy and indoor climate
An in-floor heating system has been laid into the new concrete deck floor, which can heat or cool the rooms using low-temperature heating (water < 95°F/35°C). A second system has been installed in the bedroom ceilings to facilitate additional cooling in summer.

0233 ▲

Roofs and facades have been insulated or reinsulated (R-value of 3.5). The floor has also been insulated (R-value of 3) and finished with a smooth, continuous concrete deck floor on compression-resistant insulation. The old wooden frames have been replaced with new aluminium-framed facade units of insulated glass (U-value of 1.1)

◄ 0234

Ventilation of the house is premised on the natural circulation of air throughout the building. Ventilation units in the outer walls make it possible to regulate the exact quantity of air entering the building. In summer, ventilation can be stepped up using a mechanical discharge system in the roof lights of the central hall.

0235
All living spaces receive daylight from more than one direction. Storage units, bathrooms and other ancillary spaces also receive daylight, some of it indirect. All artificial lighting is LED-based.

0236
The entire system is fed by a thermal storage unit. Self-generated energy is not being treated as an option for the time being. The surrounding trees mean that there is much shade for a large part of the year and little wind. The part of the roof that does catch the sun all year long is provided with a solar boiler for hot water (head pipe system).

0237
Rainwater on the roof is run off directly into the brook. All waste water (gray water) is run off into a tank where it is purified organically and then discharged into the brook. Only biodegradable cleansing agents are used in the house.

0238
The living room heats up quickly in winter by being oriented to the south and having all-glass facades, and it serves as a heat source for the house as a whole.

0239
The yard is watered exclusively from the brook. An electric robotic lawn mower keeps the grass at the correct height, and the plants in the garden can be tended without the need for herbicides and artificial fertilizers.

0240
Much of the furniture is built-in and, where possible, made of sustainable materials. Wood floors and wardrobes are made of bamboo; kitchen cupboards of Ecoplex (poplar) laminate; couches in the living room and kitchen, and curtains in the nurseries are made of wool felt; and floor coverings in the sunken sitting area are bamboo.

Djuric Tardio Architectes

17 Rue Ramponeau
Paris, 75020 France
Tel. +33 1 40330641
www.djuric-tardio.com

0241 ➤
A home that adapts to an urban context owing to its volume and the roof/pergola, which creates a sunny and intimate suspended garden or terrace. It is made of eco-sustainable building system with Finnish wood panels positioned on a base.

0242 ➤

A dense and pleasant urban typology, that consists of houses and small collective houses and apartments with terraces and gardens. Some of the techniques used in this sustainable district include preserving of existing vegetation and re-planting removed trees and double orientation of the homes (northeast and southeast) for better sun exposure and protection against the wind.

0243 ▼

Construction of new low-cost rental housing. We wanted to give a new image to the existing neighborhood, make the residents feel safe and create a sustainable neighborhood. The wooden constructions are in-between apartment buildings and houses. These represent a wide variety of solutions for homes, interior/exterior relationships and building forms.

0244 ▼

In Langoiran, France, an eco-district project provides environmental solutions. Wooden construction, solar panels, English gardens and various advanced technologies in energy savings were used. The urban concept is developed from different types of houses and materials. The houses are built around the principle of a common yard, a hub for social life.

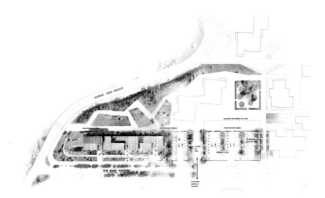

◄ 0245
Lights and shades
Hôtel à Nador, Morocco, respects the historic urban fabric and it is built around a marina, allowing direct access to the sea. The very hot weather requires a double-skin façade that generates ventilation and prevents the overheating of the walls.

0246 ➤
Moroccan national television headquarters in Rabat
The surrounding forest and its vegetation penetrate the site. This ensures the building's climate control, providing shade and cool air, creating a natural barrier and dominating the entrance to the building.

◄ 0247
Reconversion of a military barracks
The project respects the existing plaza and improves the quality of life of those around it, enriched by the social density of an urban barrier.

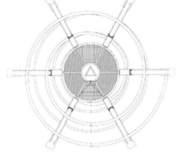

0248
**Sustainable urban,
suspended movements**
This entirely wooden suspended
path can be covered on foot or on
bike. It is a strong and dynamic
link. The plaza is an urban magnet
and a place of exchange.

0249 ➤
With its slim and atypical form, the
mobile phone antenna in the Venetian
Lagoon, becomes a metal sculpture
and light of life. The objective is to
highlight its technological aspect and
give it an additional role. Technology
is at the service of the environment.

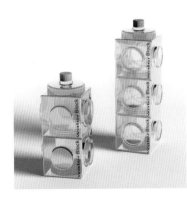

◄ **0250**
It's a bottle but not only a bottle.
Reusing plastic bottles is
environmentally-friendly.
It produces less waste.
After being used, it becomes
a multipurpose object.
It can be recycled as an object
and not only as PET plastic.
Blocks can be connected to each other.
It's a recyclable object before
becoming a recyclable material.

Dorte Mandrup Arkitekter ApS

Nørrebrogade 66 D 1.sal
København N, 2200, Denmark
Tel.: +45 33937350
www.dortemandrup.dk

0251 ➤

Think passive

The heating and lighting potential of the sun can be used in various ways, but the most efficient is the simple one: Letting the light penetrate the building in an efficient and manageable way to facilitate the functions of the building, and storing the heat of the sun in the heavy parts of the building. This ensures efficient use of the sun's energy throughout the entire life span of the building.

0252 ➤

Think active

From a holistic perspective of sustainability, the amount of time a building is actively in use directly reflects the actual need for a building. As such, a building that is not used very often can be measurably sustainable, but if it is only used very rarely, the energy used to erect and maintain the building, easily surpasses the energy saved, when the building is in use.

0253 ⏶

Why build?

Every time the building process begins, the main objective needs to be clear and well based. If the program of a building is not clear, change it. The Herstedlund Community Centre is an evolution of a well-known building typology in Denmark, but in a lot of cases that building is unused 90% of the time. In order to clarify and enhance the need for the building, the program was changed and expanded to facilitate more functions and groups.

◄ **0254**

Explore all options

The community center had a sustainability criterion added at a very late stage in the process, and the sustainable measures used were therefore based on the exploration of options at hand: soil- and sun-based heating, semi-passive ventilation and active lighting systems.

◄ **0255**

Economical accessibility

Sustainable building only matters if it is accessible to the masses. The "Platform" project is a single-family standard-house project that combines the economic benefits of mass-production with a low-emission, low-energy-consumption building.

◀ 0256

Fit in
Consider the hierarchy of the city structure before deciding which part needs to be added to the context. Add only buildings and expressions that will find their place in the context, to prevent the city structure from having to change repeatedly, resulting in an overall increase in energy consumption.

0257 ➤

Think social
The success of a building like a community center lies entirely on its ability to attract and mix as many different people as possible, as often as possible. Make sure that the building serves as an invitation, especially to users that wouldn't normally consider the possibilities of the building. This goal is only achieved by thinking all-in; thinking and building social.

0258 ▼

Densify
Non-buildable areas of the earth are a main consumer of CO_2 that would otherwise be emitted into the atmosphere. By densifying the cities as much as possible without losing the special qualities of the city spaces, our total CO_2 emissions are deducted.

0259 ▲

Early incorporation
The early adoption of a sustainable and holistic approach should be a precondition of any new building. The Platform house was done with the precondition of lowering energy consumption and emissions to 30% of the building code.

◀ 0260

Reuse
Reusing materials, especially if they are already on the site, will minimize overall energy consumption as well as add a historic layer to a new architecture. The bricks of an existing but unsalvageable schoolhouse were reused in the completion of the new school-house.

Ecospace® Ltd.

3 Iliffe Yard
London, SE17 3QA, UK
Tel.: +44 20 7703 4004
www.ecospacestudios.com

0261 ➤
Find your natural space in life.

◄ **0262**
A pitched roof is a different story
A sloping roof with skylight creates a bright and airy interior. The extra space above becomes a mezzanine chill-out pad for two-story loft-style living.

0263 ➤
It's built to modern housing standards, which means it's made to last a lifetime using the latest in modern methods of construction. We use the highest specification of structurally insulated panel systems (SIPS) to achieve exceptional structural and insulation performance, as used on many new eco homes.

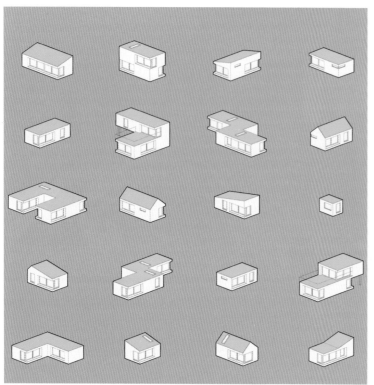

◄ **0264**
An office, a studio, a gym, an extension of your home... An Ecospace building is whatever you want it to be. Made from sustainable cedar wood with an optional plant-covered roof, as well as low-energy heating, lighting and insulation, it's right at home with the environment and your garden.

0265 ➤

It's cost effective

An Ecospace studio costs less than a home extension or loft conversion and has a fixed price, so no "unforeseen costs" from the builders.

0266 ➤

Ecospace is eco-friendly

This space has sustainable cedar wood, plant-covered roof, low-energy lighting and under-floor heating, plus high-performance insulation and double glazing.

◄ **0267**

Tailored to you

Everyone's perfect Ecospace studio is different. As an architectural practice, we can design any space required, such as one with a shower room, mini kitchen or something totally unique. We have even grown Ecospace studios into homes, holiday retreats, commercial and educational buildings.

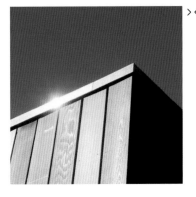

><

0268 ⋀

The unit's quick installation time—just 5 days—creates much less disruption than an extension or conversion.

➤ **0269**

Eurospace offers contemporary architecture, sleek lines and natural beauty, all at home, at the bottom of your yard.

◄ **0270**

A decade and still evolving, naturally. We were the first to develop the contemporary modular garden studios nearly 10 years ago.

Ecosistema Urbano

6 Estanislao Figueras
Madrid, 28008, Spain
Tel.: +34 915 591 601
www.ecosistemaurbano.com

◄ 0271
Develop constructive criticism in an optimistic approach to reality in order to think up creative solutions. An ECOBOULEVARD in outer Madrid, Spain, is an operation of urban recycling that reconfigures the existing urban development. Three pavilions, or artificial "trees," function like open structures to host multiple resident-selected activities.

0272 ▼
The concept of the city is completely linked to the creation of public spaces. These projects were designed to improve public spaces in central Madrid. Selected points were chosen as catalysts to initiate a broader reconfiguration of the city's public spaces.

◄ 0273
Rely on low cost to make big projects with less resources
We created a shadow space of 26,910 sq. feet (2,500 m²) to hold summer activities. The starting point is a low budget to cover the whole space using standard means.

◄ 0274
Citizens' participation and reusing standardized objects made this revitalized space possible.

0275 ▼
Build networks to share knowledge and experiences
- Our blog, ecosistemaurbano .org, is an internet platform to show worldwide experiences on creative urban sustainability.
- Euabierto.com is an open network of professionals to share experiences and knowledge on creative urban sustainability.
- Ecosistemaurbano.tv is an interactive television channel for exchanging multimedia concepts related to urban sustainability.

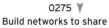

◄ 0276
Integrate the citizens into the processes of changing their environment
The Ecological Reconfiguration of Philadelphia is a strategy, based on the localization of a series of low-cost interventions on the street network, for creating mechanisms that promote citizen participation: a launching pad for the auto-regeneration of the urban fabric.

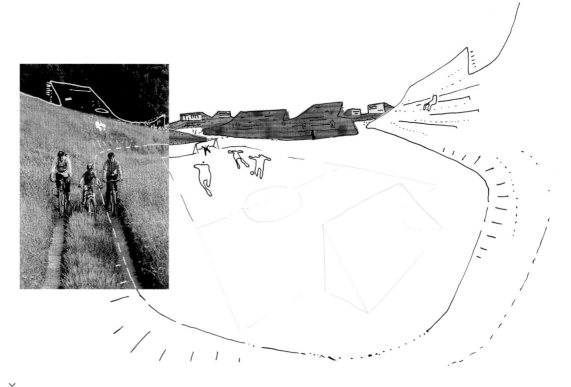

0277 ➤
Create open systems in order to allow the development of a changing reality
Maribor Landfill Urban Extension, Slovenia.
An area occupied by the municipal landfill, which is presently in a phase of ecological rehabilitation. We took this temporary circumstance as a leitmotif of the project, understanding that temporary processes should be highly related to architecture.

0278 ⋀
Reactivate the existing as an alternative to uncontrolled expansion
How to improve the public space of a neighbourhood in Madrid's outskirts? Demolish the limits that separate different urban facilities from each other and from public spaces. Reconnect these activity points and reconfigure the whole network. Consider the city as a playground.

0279 ➤
Consider the intangible by using new technologies as a mechanism to create complexity
Architecture as a means to manage energy resources (water, wind and sun) becomes the driving force for activity, forming the mutations of the landscape. A network of technological elements characterizes the image of the Waterpark to demonstrate that nature can turn any urban waste into a new resource for the city.

0280 ➤
Keep positive to be able to change reality
In contrast to traditional problem-solving methods for revitalizing degraded public spaces in historical centers, we believe another form of intervention is possible, where the citizen plays an active role in the creation of public space.

FLOAT architectural research and design/ Erin Moore

erin@floatwork.com
Tel.: +1 541 868 6074

◀ 0281
Consider the lifespan of each building material and plan for its replacement, removal or re-use.

0282 ▼
Consider the habitat that a building can offer around its exterior and make comfortable, ecologically unique places.

◀ 0283
Place window openings so they reveal different aspects of the environmental context; use near, middle and far views to include multiple levels of detail in those views.
© FLOAT

0284
Consider the physical experience of moving inside and outside and how each moment of that transition relates to a detailed sense of being in that place.
© J. Gary Tarleton

◄ **0285**
Find beautiful ways for the building to record the life around it.
© J. Gary Tarleton

0286 ►
Maximize direct solar gain during the cold season and eliminate it during the hotter months.

	MORNING (10 AM)	AFTERNOON (3 PM)
FEBRUARY	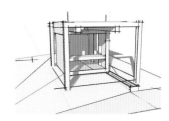	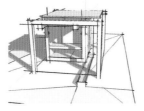
MARCH	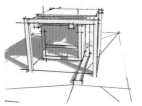	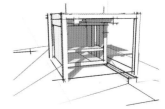
JUNE	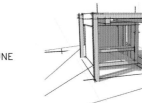	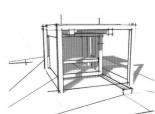
OCTOBER		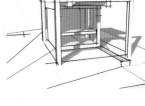
NOVEMBER	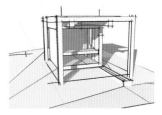	
DECEMBER	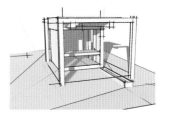	

0287 ⋀
Make the most of every drop of rain that falls on the site. Even a small basin will draw many animals.

0288 ➤
Start with the idea of putting people in a place rather than with the idea of putting a building there. Grow the building up around the people in a way that makes the most of the place.

◄ 0289
Let the structure be the building.

0290 ⋀
Attend carefully to how the building meets the ground. This connection embodies the project's relationship with its ecological context.

Francisco Portugal e Gomes,
Arquitecto

602 Rua de Vale Formoso, 2º Esq.
Oporto, 4200-510, Portugal
Tel.: +351 228 303 803
arquitecto@franciscoportugal.com

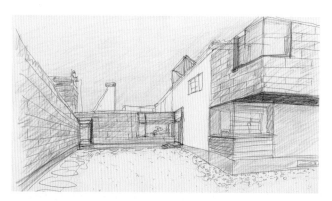

◄ 0291
For thermal protection and reduced
energy consumption, cork is used
inside and outside the building. On the
exterior surface of the extensions to
the house, the cork will be apparent
and the application will be made with
panels with high-performance thermal
insulation that has a dimension of
5 x 1 1/2 x 1/3 feet (1.5 x 0.5 x 0.1 m).

◄ 0292
The insulating characteristics of the
cork keep a constant temperature
inside, avoiding the high costs of
the energy consumption associated
with mechanical cooling systems.
Unlike other insulating synthetic
materials, cork is an environmentally
friendly material that does not
contribute to CO_2 production.

0293 ➤

The issue of external glazing in this work is one of the challenges the client made before the project started. He wanted the interior space to have a large area of glazing in order to observe the landscape that would also provide privacy for the residents.

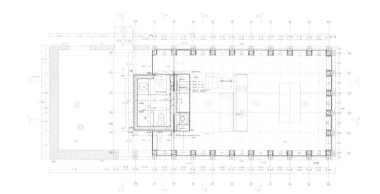

0294 ➤

Regarding solar impact, the opaque areas are formed by pillars and represent 25% of the facade's surface area, while the glass surfaces account for the other 75%. In the winter, direct gains are obtained via solar radiation, and achieved through the glazing and the application of slate on the floor, which retains the heat it absorbs.

0295 ⋀

The solution that emerged at early stages of the project was targeting the gateways traversed by the slate brise-solel. This solution, combined with the composition of the facades and the option for local materials, allowed for the implementation of energy efficiency concepts, such as controlling the solar impact and heat loss, combined with a natural ventilation system.

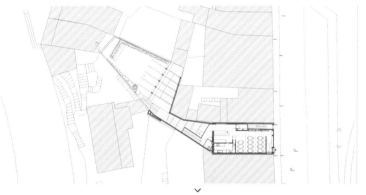

0296 ➤
The project is the full redevelopment of a small urban lot with the peculiarity of having a side lot not aligned with the building next door nor the street.

0297 ▼
The situation makes the space not built invade the back of the surrounding buildings.

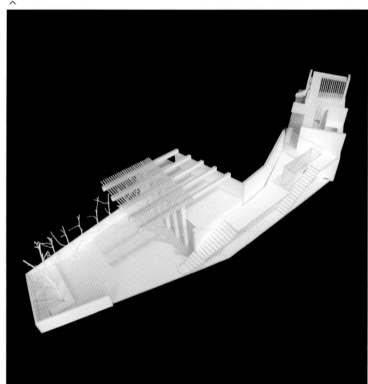

◄ 0298
The exterior space was converted into a bamboo garden. This garden, while an extension of the visual field of the tea house, also provides a contemplative function and green space on the property, which also benefits the neighbors.

0299 ➤

In Oporto, Portugal, the temperatures in July, August and September are frequently around 85ºF (30ºC). The extension has been designed with an architectural shading device that prevents the direct impact of sunlight on glass surfaces to reduce energy consumption through artificial cooling.

◄ **0300**

Another contribution to the reduction of electricity consumption was to design the interior lighting with LED (light-emitting diode) technology. Some lighting fixtures were specifically designed for this work.

François Perrin

Los Angeles, California 90026, USA
Tel.: +1 213 590 7096
www.francoisperrin.com

0301 ➤
Build with the air

A layer of air is created in between the interior wood structure and the exterior plastic skin. It acts as a natural insulation and keeps the building at the same temperature all year long. It is free, non-toxic and creates a zero carbon footprint. It is highly efficient as well (No air conditioning).

00302 ⋀
Build with the sun

A sunroom is added to a California bungalow to create a transition space in between a cold and dark interior and a hot and sunny backyard. Made of a translucent skin on top of a wood structure with several windows, a skylight and sliding doors, it brings natural light and the sea breeze into the house.

00304 ⋀
Build with local materials

Wood is one of the main construction materials in California. It is produced in the state and used for the structure, underlayer, decks, floors and exterior sidings. California cedar or redwood are perfect for the local weather. No need to deforest another part of the world.

◄ 00303
Build with the earth

The site is carved to place half of the house underground in order to protect it from the heat during the day and from the cold at night. Each floor extends outside in order to catch the sea breeze. Like the first habitations of human beings, it acts as a grotto.

0305
Build with new geometries
The volume that is designed for the house extension responds to the orientation of the site. The solar envelope is the geometrical shape that takes advantage of the sun's path for a natural energy gain through the use of passive solar heating.

0306
Build with local techniques
The wood frame is the typical construction system in the U.S. and especially in California. It is the easiest, cheapest, fastest and most flexible of any system. In this project, the structure is kept visible and only wrapped by a translucent plastic skin.
© Michael Wells

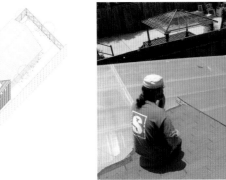

0307
Build with the wind
A light structure made of a silver aluminum mesh acts as a temporary structure to host the students of the Mountain School of Art during Art Basel Miami Beach. The shape of the structure as well as the specifics of the material allows the strong winds common at that time of the year to go through it, without impacting its stability, and cool down the temperature.

0308
Build with local people
Dante Cacace is a carpenter based in Venice Beach, California. He usually likes to work very locally, only a few blocks from the ocean (west of Lincoln). His knowledge of the local building code, materials, techniques as well as his deep commitment to the local environment are the key to many successful projects.

0309
Build with new products
Some high-tech products are built and tested for the agricultural industry, like this highly reflective silver aluminum mesh to protect plants from extreme sunlight or these translucent polycarbonate panels to cover greenhouses, and can be adapted to any project.

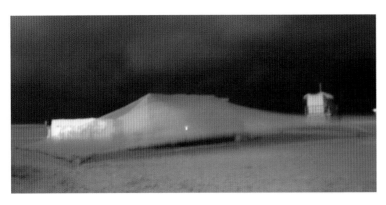

0310
Build efficiently
This is the only remaining waste of a 1,000-sq.-foot (93 m²) house prototype built entirely out of wood. The design was created according to the panel and stud dimensions in the market to avoid any extra cuts and unnecessary waste.

Garrison Architects

45 Main Street, #1026
Brooklyn, New York, 11226, USA
Tel.: +1 718 596 8300
www.garrisonarchitects.com

0311 ➤
Staten Island Animal Care Center, Staten Island, New York
Prioritize passive approaches and supplement with active systems only when required. At the Staten Island Animal Care Center, translucent multi-wall polycarbonate panels permit natural, diffuse daylight to permeate the space, making electrical lights unnecessary during daylight hours.

0312 ▼
Syracuse University School of Architecture, Syracuse, New York
Seek opportunities to introduce passive ventilation. A central atrium with an operable skylight provides a place for buoyant warm air to move upwards and out, promoting fresh air ventilation without mechanical equipment.

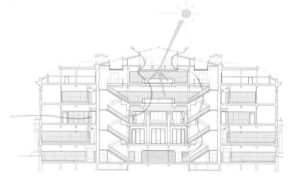

0313 ▲
Ambassador's residence, Apia, Samoa
Incorporate an understanding of climate into the earliest stages of design, and learn from the strategies of vernacular building types. Solar orientation, shading and wind breaks are effective, inexpensive strategies for reducing cooling and heating loads. Deep roof overhangs and louvered shades keep the tropical sun off the facades while promoting natural cross-ventilation.

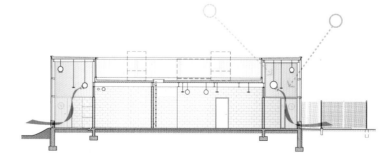

◄ 0314
Staten Island Animal Care Center, New York, New York
Recycle heat energy. Using a heat recovery wheel in the ventilation system exhausts indoor pollutants from the air without losing heat. In cooling season, the system reverses, transferring excess heat in the ventilation air to the cooler exhaust air, reducing the need for mechanical cooling.

◀ 0315
Red Hook Green, Brooklyn, NY
Realize net-zero energy by connecting rooftop photovoltaic arrays to the municipal power supply grid. Red Hook Green sells back excess solar energy from sunny days to the grid, maintaining reliable backup power for rainy days. At the end of the year the overall energy production meets or exceeds the building's consumption.

◀ 0316
Koby Cottage, Albion, Michigan
Consider the factory. Buildings constructed off-site can recognize huge efficiencies by consolidating time and resources. Modular designs are driven by the efficient use of space and materials, resulting in buildings that can be incredibly functional without sprawling.

0317 ▼
Lehman College Child Care Center, Bronx, New York
Integrate plantings into vertical surfaces. Plants can be grown in layered facades with shading devices or rainscreen cladding. For low-maintenance installations, thermostatic control of operable facades and hydroponic watering systems maintain optimal conditions for plant health.

0318 ▲
Red Hook Green, Brooklyn, New York
Think beyond the photovoltaic panel. Highly efficient technologies to harness the sun's thermal energy have been developed for applications ranging from heating and electricity production to cooking and distilling drinking water. The Red Hook Green project uses high-efficiency solar collectors to provide its domestic hot water.

0319 ▲
Border Patrol Station, Murietta, California
Take advantage of natural daylight to reduce electrical lighting costs. Use light shelves to bounce daylight off of building surfaces, preventing glare and creating even, diffuse ambient lighting. Similar techniques, such as cove lighting, can also be implemented to increase the effectiveness of artificial lights.

0320 ➤
Border Patrol Station, Murietta, California
Hybridize. Use natural ventilation in circulation and common spaces while providing active cooling dedicated to enclosed spaces. The spill air from the enclosed spaces can be used to temper spaces that can accept higher temperatures.

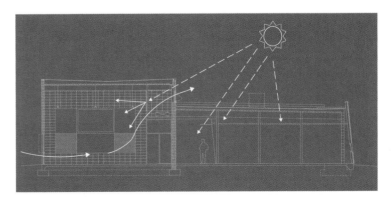

Green Dwellings / Leger Wanaselja Architecture

2320 McGee Avenue
Berkeley, California 94703, USA
Tel.: +1 510 848 8901
www.greendwellings.com

0321 ➤
Old aluminum highway signs, attached from the outside in, make a durable siding on this bay window. Reusing materials before they are recycled can save significant energy and reduce the environmental impacts of processing. For this project, old highway signs were also bent for light shades and stair railings.

0323 ▼
Using locally sourced junk and recycled materials can significantly cut back on environmental impacts. This green renovation and new construction project that features small, energy- and resource efficient- units also relies on extensive salvage and reuse, including 3 1/2 tons of discarded street signs.

◄ **0322**
Junked streets signs are incorporated into this pedestrian fence and gate. A Volkwagen headlight lights the entry at night.

0324 ➤
The upper level of this house is clad in over 100 gray car roofs. The roofs were carefully selected for color and sawn out in the junk yard. This new, two-bedroom, 1,140-sq.-foot (106 m²), virtually passive solar house is an infill in one of Berkeley's oldest neighborhoods.

0325 ➤

Tempered car windows are versatile as building components because of their size, three-dimensional shape and mounting hardware. Here, Mazda hatchback glass is repurposed as a railing. Far right shows a Porsche rear window used as an awning.

0326 ▼

Wood and wood products remain among the most environmentally friendly and versatile materials we use in construction. The lower walls are clad in poplar bark, a waste product from the furniture industry of North Carolina. FSC-certified wood was also used for framing. And, the walls and ceilings are insulated with shredded old phonebooks, newspapers and cardboard.

◄ 0327 ▲

Built to exacting specifications to withstand tremendous loads and, with refrigerated units, to maintain low internal temperatures efficiently, shipping containers make excellent building blocks for architecture. Even after many trips around the globe, they still easily meet or exceed the building codes for construction. As a net importer, the U.S. ports are filled with them. Here, three insulated containers are used to create this airy yet compact 1,330-sq.-foot (124 m²), three-bedroom, house.

◄ 0328

The awnings are fabricated from junked Dodge Caravan side windows. Once advertised as "America's best-selling minivan," they are now a common item in junkyards.

0330 ▲

Two, 200-year old oak trees fell on the property during the design phase. Not letting their demise go to waste, the trees were extensively incorporated into the design of this remodeled house. In fact, almost all of the finish wood in the project, including fencing, siding, cabinetry, and flooring, is salvaged, totaling over 6,000 board feet (16 m³) of wood.

0329 ➤

Other car parts also perform well in construction. Here, hatchbacks were used to make a railing. Elsewhere, eight Volvo rear doors are fashioned into a fence and a remotely operated electric parking lot gate.

GROUP A Architects

3 Pelgrimsstraat
Rotterdam, 3029 BH, The Netherlands
Tel.: +31 10 244 01 93
www.groupa.nl
www.groupalive.com

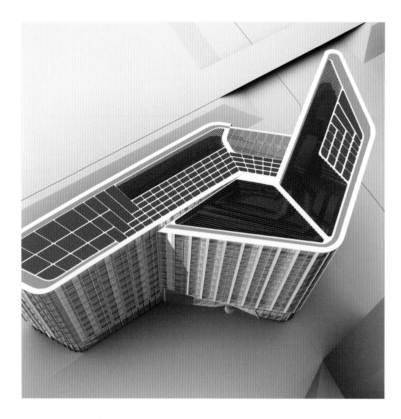

0331 ▼

Do the Rs

The classic Rs can't of course be missed: reduce, reuse, recycle. When reducing, savings are both economic and related to sustainability. Explore reduction before deploying reuse and recycling. Reuse (adaptive reuse) can happen on a big scale and often results in significant savings and environmental benefits, but it takes planning. All selected building materials, components and parts, should be recyclable.

Aviodome for SALE

0332 ▲

Do think ahead

Essential to the early stages of planning and designing a green build is to think ahead to environmental impact, orientation, existing surrounding qualities, technical practicalities, social and cultural cohesion, selected materials and finishes, the design and construction process, the long-term sustainable "value" maintenance, education of end-users, and future change in expectations and demands.

◀ 0333
Go "new world of work"

In the early stages of design, consider the space's end users and what they require. Discuss, for example, the office's occupancy rates, the potential of multi-functional spaces, the value of conference tables and open-plan activities. Develop a more compact office environment with less fixed workspaces, anticipating realistic occupancy rates. Prepare for the new world of work.

0334 ▶
Go compact

Once building activities are underway, and all stages of reflection and appropriate sustainable considerations have been completed, it is an additional "green" asset to build as compact as possible. Smart urban growth, clever higher densities, expanding vertically rather than horizontally, minimizing circulation distances, atrium-embracing floors, and new world of work concepts, all contribute to this ambition.

0335 ➤

Do reconsider

Before considering any physical preparation and building activities on the site, parties involved in the planning should make sure that the building is well considered. It should be absolutely clear to everyone that the envisioned build has social, cultural, environmental, and economical relevance, now and in the future. Reconsidering any aspect of the build that isn't working should always be part of the planning process.

◄ **0336**

Be inspired

Acting "eco" can only be considered succesfull if the final product fits the user, fullfills his expectations, and creates an atmoshere that motivates and inspires him. Reference to this objective in GROUP A designs, is our own studio. Although the 1930 monument still has shortcomings, the building has contributed greatly to the creative process and solid social cohesion amongst staff.

0337 ▲

Do love

Unconditional love for our profession – designer, creative architect, educator, communicator, value adder, sociocultural worker, urban planner, motivator, initiator, team member – is the ultimate motivation in the way we live up to our principles. Only with this mentality and passion can we start inspiring other people and processes to be eco-aware, and ready to implement 1,000 eco tips.

◅ 0338
Be natural
It is vital to consider how build environment will impact people's physical and mental well-being. The natural quality and performance of building materials, internal climate, daylight and interaction with the outside are key to the desire to provide a healthy, sustainable build. Choices should therefore be inspired by the ultimate balance and principles, as they apply in nature.

0339 ➤
Be receptive
Planning, designing and building is teamwork, its success often depending on the willingness of parties being receptive to each others' opinions. Open up and get inspired by cooperating disciplines. Start listening more with your heart and your intuition, instead of your mind. Get rid of the habit of "being heard" to be able to really contribute to innovation and sustainability.

◅ 0340
Go underground
By deploying relatively low-tech means, building performance can improve substantially. For example, exposed internal concrete structures inhibit temperature fluctuations. The natural steady temperature reduces energy consumption and requires smaller plant facilities. But wouldn't it be even more effective to minimize the external impact of extremes in temperature altogether by going underground and introducing green roofs?

Gudmundur Jonsson
Arkitektkontor

24 Hegdehaugsveien
Oslo, 0352, Norway
Tel.: +47 23 20 23 50
www.gudmundurjonsson.no

0341 ➤
The layout is precisely organized according to the most important views, giving glazed fronts toward the view and closing the building as much as possible otherwise to emphasize the idea.

0342 ▼
Thus, the choice of materials becomes important , both so that they match the landscape and to ensure they are sustainable.

◄ **0343**
The materials used are concrete (with a natural pattern), zebra-wood (very strong and durable) and Icelandic bluestone.

◄ **0344**
Glazed openings were prioritized in order to emphasize the most important view, and the rest of the building was closed in order to increase expectations for the view that can be experienced.

0345 ➤
Use roofs to recreate the landscape, using it for recreation and to insulate the buildings below.

><

0347 ➤
Slate is a product leftover from the production of larger stones.

0349 ➤
Materials are related to the climate of the site in order to be climate-resistant while simultaneously requiring a minimum of maintenance.

0346 ⬆
The materials are practically maintenance free.

0348 ➤
Using natural materials, such as slate, and durable slow-growing wood such as Siberian wood *(Larix sibirica)* creates an energy-conserving and age-resistant building.

◀ **0350**
Making architecture by interpreting the culture of the place and using the elements consciously, enhances the energy-conserving building parts.

Guz Architects

Singapore, 278199, Singapore
Tel.: +65 6476 6110
www.guzarchitects.com

0351 ➤
Keep built forms to a minimum to maximize space for nature/greenery.

0352 ▼
The smaller, more efficient a house is, the lower its carbon footprint will be.

0353 ▼
Lay out the building to make the most of the surrounding environment. Don't destroy it, but keep the "magic" as much as possible.

◄ **0354**
Design for passive cooling/ heating as much as possible.

0355 ➤
Use locally available materials where possible; cut down on imported materials from far away.

0356 ▽
Use sustainably formed timber if possible as a carbon store – it has a much smaller footprint.

0357 ▲
Recycle water and use sunshine for photovoltaics where possible.

0358 ▽
Green roofs create more greenery, which is easy on the eye and keeps a house cool.

◄ **0359**
Keep to a human scale – overly large buildings don't look good and don't relate to nature.

0360 ▽
Bring light into the basement.

Harquitectes

22 Carrer de Montserrat, 2n 2a
Sabadell, 08201, Spain
Tel.: +34 93 725 00 48
www.harquitectes.com

0361 ➤
Our design approach is based on common sense and is closer to a radical normality than strident exhibitionism.

0362 ▼
We are interested in popular architecture, and, most of all, in the technical conditions necessary for its construction to make it a formal and technical reference.

◄ **0363**
We believe in sustainability, not as a new paradigm or as a single or sufficient objective, just as a more professional framework.

0364 ▼
We want to become experts in all available materials. Make the best use of them and understand their characteristics. Focus on the constructive logic peculiar to them, or find new ones. Get the most out of them in terms of both material and space.

0365 ▲
Architecture requires a total turnaround. It needs to be redefined. We must listen to the social demand (real) and try to anticipate needs. As architects we should try to lead this social redirection, especially regarding the territories that are more typical of our profession and that, in theory, we master better than anyone.

0366 ➤

We have an obsession with a concept we call space-structure, and we are always committed to make the most of the materials we use. Casa 205 perfectly exemplifies both concepts (space and material).

0367 ⬆

We review vernacular architecture and we name it available architecture: available materials, available techniques, available typologies. If we understand vernacular in this way, nothing prevents us from imagining the existence of a modern vernacular construction. Today, the availability is not based on geographic terms but economic terms. Construction should be catalog, prefabricated, lightweight, systematic: using market systems that are assembled rather than built. Screwing instead of hitting.

◄ **0368**
We opt for those strategies that achieve the highest architectural and environmental quality as well as simplifying, reducing or reinvesting costs in order to provide greater living comfort and greater management of the building.

0369 ▼
Incorporating low-energy biospheric materials, designing passive strategies to reduce energy demands and enforcing both reversible and reusable building systems seem consistent with both available and sustainable construction.

Summer

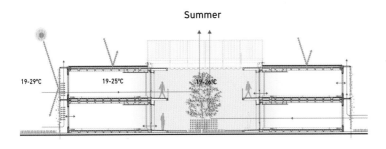

19-29°C   19-25°C   19-26°C

Winter

0-13°C   18-20°C   10-20°C

◄ **0370**
We take sustainable architecture seriously, starting from the understanding of the life cycle of the building: materials, construction, energy demand and construction.

H2o Architects

29 Northumberland Street
Collingwood, Victoria, 3066, Australia
Tel.: +61 3 9417 0900
www.h2oarchitects.com.au

0371 ➤

Brise-soleil
To the north elevation of the same building, 10-foot (3 m) wide balconies double as a sun screen device while providing access to internal academic spaces as a substitute for an internal corridor. Use of open "pores" in the precast facade provides borrowed light to internal rooms. (Swinburne University of Technology, Advanced Technologies Building, Hawthorn, Victoria, Australia.)

0372 ▼

Recycled materials
Facing bricks from a former building demolished by the university prior to construction were stored off-site and reused extensively on the balcony faces of internal areas to maintain thermal performance and provide a weather-proof, "urban," low-maintenance finish. (Swinburne University of Technology, Advanced Technologies Building, Hawthorn, Victoria, Australia.)

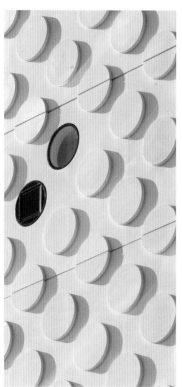

◄ 0373

Thermal mass
Universal application of precast concrete panelized facade adapts flexibly to a variety of user conditions. Combined with minimum acceptable glazing to satisfy briefed requirements and limited glazing to the east and west this ensures high thermal mass, (Swinburne University of Technology, Advanced Technologies Building.)

0374 ▲

Off-site construction
Off-site factory manufacturing under controlled factory conditions was used for all envelope components including structural framing, walling, roof trusses and roofing, seating platforms and seats. (North Stand, Lakeside Stadium, Albert Park, Victoria.)

0375
Adaptable reuse
An existing 19th-century rope factory in inner Melbourne was simply converted to a state-of-the-art IT "barn." Acoustically and environmentally contained pods with a grid "ceiling" act as a carrier for introduced services. (RMIT University, Building 512 Computer Barn, Brunswick, Victoria, Australia.)

0376
Screening device with multiple uses
Folded and undulating timber screen in celery top pine (unique to Tasmania) acts as a trellis for a heritage grape vine and provides sun protection, visual privacy from an adjacent public footpath and, by extending 4 feet (1.2 m) above an upper roof terrace, acts as a handrail. (Hobart Tennis Club, Hobart, Tasmania, Australia.)

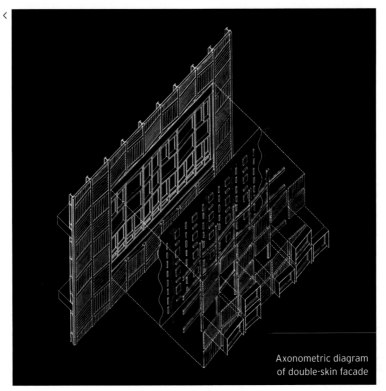

Axonometric diagram of double-skin facade

0377
Rainscreen
A textured Western red cedar panelized facade recalls textile analogies and acts as a rain screen in front of the Tyvek-sealed building envelope behind. A 3-inch (7.5 cm) air gap separates the two "skins" and aids the partial cooling of the north facade through an induced thermal chimney effect. (RMIT University, Building 513 [Textiles, Clothing, Footwear & Leather], Brunswick, Victoria, Australia.)

0378 ➤
Glare and sun minimization
Use of a classical "saw tooth" south light roof profile as sun control device to a library that was briefed to have minimum glazing. (Avondale Heights Library & Learning Centre, Moonee Ponds, Victoria, Australia.)

⌄
⌃

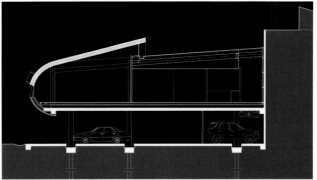

0379 ▲
Cross-ventilation
Establishing a sectional profile to a linear, extrapolated plan maximizes cross-ventilation. Mechanically operated (via BAS system), low-level labyrinth style vents to the south side and high level louver vents to north side were used to maximize clean air intake.
(State Emergency Services Headquarters, Southbank, Victoria, Australia.)

◀ 0380
Thermal chimney
Internal multi-level atriums along the main spine of a large educational center act as thermal "chimneys" to extract hot air at night via mechanically operated high-level glass louvers that are controlled by a BAS system.
(Deakin University International Centre & Business Building, Burwood Campus, Victoria, Australia.)

Heliotrope Architects

5140 Ballard Ave Northwest, B
Seattle, Washington 98107, USA
Tel.: +1 206 297-0442
www.heliotropearchitects.com

0381 ➤

Energy: Envelope

A high-performance thermal envelope is the critical first step in the development of a low-carbon building. The Passive House performance standard is perhaps the most rigorous in use today. In our first Passive House-compliant project, we are utilizing a super-insulated, nearly air-tight envelope with high-performance glazing in order to drastically reduce energy requirements for heating.

0382 ▼

Energy: Inputs

Leveraging free energy provided by nature is another critical component in a low-carbon building. Thoughtfully placed glazing can drastically reduce necessary supplemental heat, and thoughtfully designed openings can eliminate the need for air-conditioning in mild climates. In our Passive House project, solar heat gain is provided through an expansive skylight with unobstructed southern exposure.

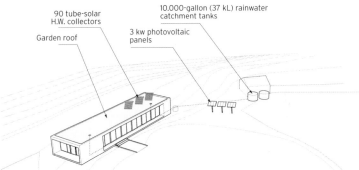

90 tube-solar
H.W. collectors

10.000-gallon (37 kL) rainwater
catchment tanks

Garden roof

3 kw photovoltaic
panels

0383 ▲
Energy: Systems
Use of appropriate technology is an
important tool in reducing a building's
carbon footprint. In our North Beach
project, a solar hot water system
provides inputs to both potable hot
water and radiant space heating
systems, a rain-water catchment
system cuts down on potable water
use and a PV system provides
supplemental electricity. These systems
have varying pay-back dependent
upon climate and local incentives.

0384 ▲
Ecology: Water
Much of our work is on or near the shoreline of the Salish Sea, which is imperiled by the cumulative effects of thoughtless development. It is important to understand and respect natural surface and sub-surface water flows in order to prevent further damage to our environment. Our Suncrest residence utilizes garden roofs, dispersion trenches and pond replenishment to filter all rainwater captured by building roofs.

0385 ▲
Ecology: Resource use
Over a 50-year period, carbon utilization in building renovation and repair has roughly equaled that of initial construction. Proper control of moisture, use of durable materials and thoughtful detailing are important components to reducing this statistic. Equally important is creating aesthetically timeless works.

0386 ➤
Ecology: Flora and fauna
In the Salish Sea, our endangered orca feed exclusively on salmon, which feed upon forage fish, which feed upon insects, which in turn depend upon a healthy near-shore habitat. The interconnections are vast, and it is critical to understand them in order to minimize harmful impacts.

◄ 0387
Experience: Light
In our region, we experience over 200 days of cloudy weather per year. Daylighting in this climate is a critical component to mental health and wellbeing of both individuals and community. We desire a luminous quality in our work and carefully placed fenestration to bathe wall surfaces in light, and always from more than one side of a room.

0388 ▼
Experience: Place
Imagine someone blindfolded and brought to a project from far away. When looking around, do they have a sense of being in a unique region and place? How is that uniqueness conveyed? Beyond views and landscape, the interrelationship of building and landscape as well as materiality, color, quality of light, etc., all work together in creating a unique sense of place.

0389 ►
Experience: Connection to environment
Establishing a strong connection between building inhabitants and the surrounding environment is central to our work. We seek to celebrate the best features of the site while also highlighting the overlooked smaller moments – a tree stump or an interesting rock can become a focal point of a more intimate space. Identify opportunities to provide connections and exploit them to the extent possible.

0390 ►
Experience: Environmental health
Fresh dust- and bug-free air, generous daylight, clean water – these are critical elements to a healthy environment. The search for healthier and more sustainable systems, materials and coatings is a never-ending process. We have spent years accumulating a lengthy list of manufacturers and suppliers to draw from, and the search is a part of our regular daily routine.

Hiroshi Nakamura & NAP Architects

2-15-7-1F Sakurashinmachi
Setagaya-ku, Tokyo, 154-0015, Japan
Tel: +81 (0)3 5426 0105
www.nakam.info

◀ 0391
The building's form wraps
around the trees.

0392 ▼
Wood columns and beams were used
since they can be easily worked to
match the complicated shapes of
the walls and ceiling, and structural
plywood was applied to create a
monocoque structure. FG board
(strengthened with inorganic fibers)
was used on the inside to follow the
many curves, and it was finished
with an elastic coating material.

0393 ▲
Specially ordered asphalt shingles
that also follow the curves and
still have a good appearance when
littered with leaves were applied to
the outside. The trees were planted
according to the plan formulated
when the building was designed after
all other work was completed.

◀ 0394
**Bring nature, buildings and
people closer together**
There is a small private art museum
in the city of Oyama, in Tochigi
prefecture, Japan. The owner
wanted to build a room to showcase
paintings that were collected by his
late father, Roku Tsukada, and a café
where people can drop in anytime
that has the ambiance of a salon.

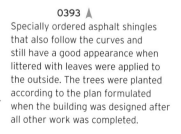

0395 ▼

For the roof slab, we used soil from the site, which provides a protective layer. To prevent any runoff, we planted wild grass seeds into the soil.

◄ 0396

This grass-and-soil roof has an insulating effect and reduces the cost of the exterior sheeting. Using soil from the site reduced the cost of transportation. The exterior walls also needed to be coated with something to prevent damage from salt. We mixed the soil, cement and resin and spread it thickly on the roof to maximum thickness of 2 inches (5.5 cm).

0397 ▲

The roofline and colors are determined by the plants. After construction is completed, the roofline and colors will continue to change by the seeds birds or winds carry, the pruning by the resident, weathering and the seasons. This house will never be truly completed.

0398 ▲

This large room gets a view of and winds from the sea to the mountains, insulating the eyes from the houses on either side.

0399 ➤

Working with the soil in a garden is a conversation the owner has with the soil depending on seasonal changes. So, when considering garden architecture, the owner and nature, through the climate and the earth, have decisive power.

0400 ➤

As finishing is being completed, clients have also joined in the work. They scratched off the soil freely with various tools such as a metal skewer, metal broom, and spatula.

HOK

211 North Broadway, Suite 700
St. Louis, Missouri 63102, USA
Tel.: +1 314 754 4215
www.hok.com

0401 ▼

Use what's free
A fundamental sustainable strategy
is to use resources that are
available and in abundance. Just
as nature doesn't import energy
and materials from distances, we
need to embrace our conditions
and seek innovative solutions.
The Project Haiti replacement
orphanage plan is strategically designed
to maximize natural ventilation by
harnessing eastern trade winds.

◄ 0402

Look to nature for solutions
Nature's survival strategies can be
directly adapted to design at all scales.
Using biomimicry, we need to identify
a design's functional challenges and
then look to nature for solutions.
With Project Haiti, the variegated
bamboo screens surrounding the walls
function like bark on a tree, protecting
the building by reducing heat, providing
shade and maximizing airflow.

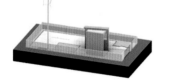

0403 ➤

Build resilience through diversity
"Diverse structures form in response
to abiotic disturbances, resulting
in an overall structure that creates
a more resilient system."
  To ensure continuity in all conditions,
the design of the Project Haiti
replacement orphanage incorporates
multiple sources for energy generation,
including wind generation, natural
ventilation, photovoltaics and a
biofuel generator for load balancing.

◄ 0404

Architecture without architects
Bernard Rudofsky said, "Vernacular
architecture does not go through
fashion cycles. It is nearly immutable,
indeed, unimprovable, since it serves
its purpose to perfection." He got it
right; there's so much to learn from
understanding vernacular solutions,
and nature's principles, in seeking
high-performance, inspirational,
lasting design solutions.

◄ 0405
Passive solutions first
Use every means at your disposal to drive down energy and water loads before integrating systems solutions. This starts with early design analysis to optimize orientation, massing, openings, building envelope and renewable energy generation.

0406 ▲
Failure is multiplicative
Performance-driven design requires all strategies to work together in harmony. If the natural lighting strategy doesn't work, the lights will stay on longer, the increased heat gain will cause the MEP systems to not perform as designed and the total loads will increase. The Net Zero Court prototype's design uses espaliered trees that provide shade but do not interfere with natural light penetration.

0407 ▲
Daylight rules
Optimization of natural light is a major design driver. This single goal brings a number of benefits, including reduced electricity usage, decreased greenhouse gas emissions, reduced heat gain from lighting and improved indoor environment for occupants. For Process Zero, a zero-footprint retrofit of an existing office building, the design team "carved" the city-block-long building to bring in natural light.

0409 ▲
Embrace your limits
Don't see limits as roadblocks to get around, but rather as ways to force an entirely new approach to developing creative solutions. After all, "necessity is the mother of invention." It's how we leapfrog to entirely new solutions. Process Zero's breakthrough design idea uses energy-producing microalgae to help power the building by capturing carbon monoxide emissions from the nearby Santa Ana Highway.

0408 ▼
There is no "silver bullet" solution
There is no simple solution in sustainability; ideas must be integrated into all aspects of the design. Every opportunity to elegantly optimize performance and environmental, social and economic benefits should be employed.

0410 ➤
Be of place
A sustainable project should tell us about its place. Like in nature, it should be locally attuned and responsive. The building orientation and massing, fundamental energy and water strategies, types of materials and design aesthetic should all tell a story about the specific climate, habitat, culture and people of a place.

House + House Architects

1499 Washington Street
San Francisco, California 94109, USA
Tel.: +1 415 474 2112
www.houseandhouse.com

0411 ▼
Small footprint
Careful analysis of rooms and their relationships to each other can create efficient circulation without the need for hallways. By rethinking spaces that are often separate, for example, family and living, it is possible to design beautiful combined environments that don't waste materials or systems.

0412 ➤
Accessibility
Making accessibility an integral part of design, rather than a special added component that must be legislated and prescribed, is a more holistic approach to creating space. This 165-foot (50 m) long ramp provides universal access and is experienced as a sculptural, three-dimensional movement through the garden, with the views unfolding as a surprise.

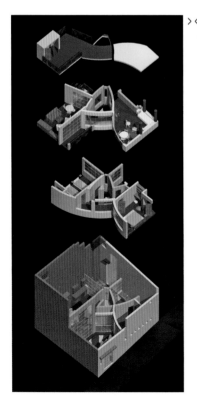

◀ 0413

Solar shading

Trellis shading offers the opportunity to create beautiful moving shadow patterns that surprise and delight the eye while offering relief from unwanted sun intrusion. A careful analysis of sun angles can provide the criteria required to allow the sun in during winter while blocking it in summer.

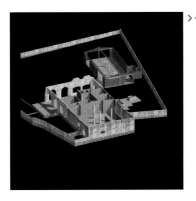

> <

ˇ
ˇ

0414 ▲

Adaptive reuse

Adaptive reuse is a good way of keeping the massive amount of debris created when a building is torn down out of landfills. This abandoned factory was transformed into a gracious home and art studio. Debris from demolished walls was used in the foundations.

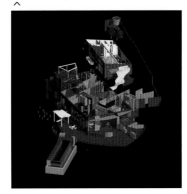

0416 ▶

Furniture

Use old materials in unexpected new ways. In this example, an old, deteriorated carpenter's work table is reframed and converted into a new dining room table that is a conversational centerpiece and invites the users to touch the markings of work done generations before.

◀ 0415

Recycling

Using recycled materials can add warmth, texture and a bit of history to new, modern buildings. An old barn was transformed into beautiful wood flooring in this home in northern California.

0417 ➤

Courtyard

The central courtyard is a vernacular concept that developed in almost every country – allowing light and fresh air into each room and inviting the inhabitants to partake of protected outdoor space. It offers the opportunity for large windows without the need for privacy screening and frames views from one part of the house to another.

◄ **0418**

Natural light

Make the sun work for you. Creative positioning of skylights and windows can provide natural day lighting to give a soft glow, make a deliberate streak or add a dappled texture. Capture its heat in the winter months in a concrete floor to re-radiate its warmth on chilly evenings.
© House + House Architects

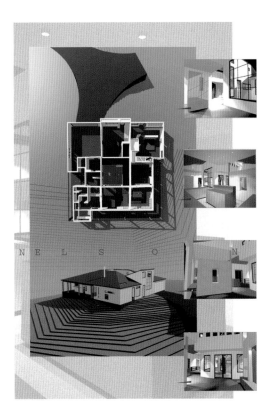

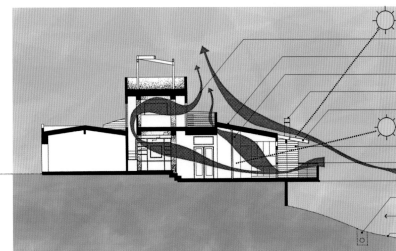

West sun – summer
Natural chimney cooling at tower
Sloping roof to deflect wind
Natural cooling with high windows
Fireplace with outside air source
Trellis
West sun – winter
Wood siding at walls with R-19 insulation
Double-glazed windows
Gale force winds
Wood flooring with R-30 insulation
Septic field
Well, water and propane tanks
Natural grade

0419 ▲
Sun control
Know where the sun is and how to really access it to your advantage. This home in California closes down at its outer perimeter to deflect the summer heat, but it opens huge window walls to a small central courtyard, where the sun has a carefully controlled entry.
© House + House Architects

◄ 0420
Convection
Use convection. Hot air rises no matter where you are or what the budget is, so make it work to ventilate your home naturally. Incorporate high operable windows to let the warm air out and low windows in shady, protected areas to let cool air in to ventilate the house naturally.
© House + House Architects

Huttunen-Lipasti-Pakkanen Oy, Ltd.

Iso Roobertinkatu 41 LH8
Helsinki, 00120, Finland
Tel.: +358 9 694 7724
www.h-l-p.fi

0421 ➤
Everything starts with the site
The biggest decisions in the energy consumption/gain of the building are made when placing the building and giving it its form. Every design should be a result of a thorough site analysis.

0422 ▼
Places can be built with gestures
The work of an architect has a lot to do with articulating spaces for functions. It is not always needed to build a heated and insulated room to create a defined space.

0423 ▲
Architecture grows from the chosen materials. All the decisions, from large scale to the smallest details, should follow the choices of materials and the way they need to be treated. The materials should turn old in a beautiful way to prevent a constant need for renovations.

◄ 0424
Buildings should be designed considering that the users and functions will change. The more flexible a plan is, the better the building's chances are for a long survival. In the case of this ice skating cafeteria, the building is designed to be divided into two pieces and moved between a beach and an ice skating arena twice a year.

0425 ▾
To build any building it takes an enormous amount of natural resources. For this reason, they should be built to last for generations. All the aspects of working with architecture are involved in this. The following points are there to illustrate this.

0426 ➤
Architecture must grow from the site and the local culture. If a building does not manage to gain the acceptance and pride of local people, it will not be taken care of and will not last for a long time. In the best case, the building is a fresh interpretation of its surroundings and a continuation of the whole built culture.

0427 ▾
Things that have quality in them are to be preserved. Whether it's a city or a door handle, well-made things survive the challenges of both new ideologies and being used over decades. The quality is defined in taking the detailing all the way to the very end.

0428 ➤
If a building is built to be temporary, it should be considered very carefully how it is going to be recycled after its use. In this day care center, the concept consists of movable warm core modules and cold facade structures that can also be undetached and transported inside the modules.

◄ **0429**
Architecture grows from the structure. The chosen structure defines the architectural language. To minimize the use of building materials, it is worth considering if the structure can be shown as it is. Clear, simple, beautiful and understandable buildings don't go out of fashion.

◄ **0430**
Every building should be unique. It is not necessary to spend vast amounts of money, energy and materials to reach this. Simple ideas are sometimes enough. The aim is to get a big effect from a little change.

I-Beam Design

138 Spring Street, 2nd Floor
New York, New York 10012, USA
Tel.: +1 212 244 7596
www.i-beamdesign.com

◄ 0431
Reuse
The ascending grassy knolls accommodate urban residents with public gardens and private lawns at their doorstep, even for people living in a fourth-floor duplex. The grass roofs provide water storage during periods of high rainfall, reducing runoff to the local drainage system, and the rain water that percolates through the roof can be filtered and used on-site.

0432 ➤
Regenerate
Green Springs help to absorb carbon dioxide, release oxygen and mitigate the presence of greenhouse gases in the atmosphere. They reduce energy consumption by keeping the building cooler during summer and minimizing "heat island" effect.

0433 ▲
Integrate
Boating is the most energy-efficient means of transportation. The apartments along the canal capitalize on this by featuring a barge summer home that plugs into the apartment, extending the living space into a floating sunroom with extra bedroom, kitchen and bath.

0434 ▼
Enjoy nature
Water is synonymous with health and life. Swimming, ice skating, and floating performances are just some of the activities that could take place in the Water Plaza at the center of this housing development.

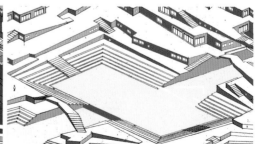

0435
Reappropriate

The principal building module for the refugee housing project is the wooden shipping pallet, which is widely available in most countries, versatile, recyclable, sustainable, easily assembled and inexpensive.

0436
Adapt and change

The pallet matrix permits families to build according to their own needs with local resources, thereby fostering empowerment, employment, and economic growth within the community. A simple pallet structure evolves naturally from emergency shelter to permanent house with the addition of indigenous insulating or cladding materials, like rubble, stone, earth, mud, plaster and concrete.

0437
Recycle

40 million bed springs are dumped into American landfills, yearly. By combining two rapidly growing industries, green wall systems and mattress recycling, Green Springs intends to reduce landfill waste and encourage the beautification of urban landscapes by using discarded bed springs to create green walls and roofs.

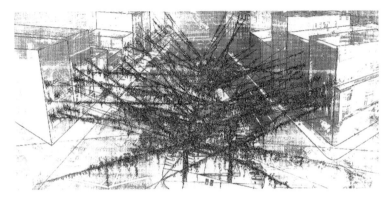

0438
Remove and relocate

By combining pallets with local construction methods, such as thatch roofing or wattle and daub, the Pallet House provides a hybrid solution that is prefabricated, environmentally sustainable and culturally appropriate.

0439
Connect

This is a living park, transformed each season by the imagination of its users. Residents participate directly from their windows by selecting, growing and planting the flowering vines and ivy that constitute the hanging garden. Seating, playing and entertainment are integrated into a playful landscape of paving, planting and water that inspires joy and interaction between people and the parkscape.

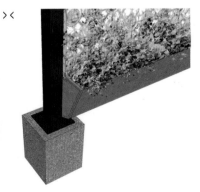

0440
Enliven

By using the sky plane as a garden, the design reduces noise and air pollution to the surrounding residences, animates the park, and provides shade.

Iredale Pedersen Hook
Architects

Murray Mews, 329-331 Murray Street, Suites 5 & 6,
Perth, Western Australia, 6000, Australia
PO Box 442 Leederville, Western
Australia 6903, Australia
Tel.: +61 8 9322 9750
www.iredalepedersenhook.com

0441 ➤
Walmajarri Community Centre
Culturally responsive architecture
seeks to bind ancient culture to
place while helping to preserve a
rapidly disappearing language.

0442 ▽
Tjuntjuntjarra Community Housing
Community architecture should be
designed with flexible individual-
use patterns while encouraging a
connection to the community.

0443 ▲
Sheep House
A heavily grounded architecture with
a minimum carbon footprint that
produces power and stores water.

0444 ▲
Swan Street Residence
Reinterpret older residential buildings with an extroverted quality that responds to contemporary ideas for dwelling. The role of the plasma screen is challenged by a dynamic engagement of interior and exterior.

0445 ➤
Perth Zoo orangutan enclosure
Assisting with the preservation of a species while responding to unfamiliar patterns of use (tree dwelling).

◄ 0446
Gidgegannup
A delicate line in the landscape that
minimizes impact to the place.

◄ 0447
Florida beach house
Intensifying the relationship of the
occupant to the ocean and challenging
building design guidelines.

◄ 0448
Meares Residence
Redefining society's understanding of the aesthetics of a sustainable dwelling. A sequence of pavilions are orchestrated on an abstract concrete platform.

0449 ▼
Information Pods
A brief moment of clarity and legibility in a context of visual chaos. Extra-small projects can have a large impact.

◄ 0450
Innocent Bystander Winery
A shifting scale of experience to engage with a small scale town, exposing the wine making process, "reconstructing" the vineyard plantations (in paint) and the texture and fine grain of recycled jarrah timber.

J. Mayer H. Architects

54 Bleibtreustrasse
Berlin, 10623, Germany
Tel.: +49 30 644 90 77 00
www.jmayerh.de

0451 ➤
Property development group
Euroboden is building a
unique apartment house at
Johannisstraße in Mitte, Germany,
Berlin's downtown district.

0452 ▼
Sonnenhof consists of four
new buildings with office
and residential spaces.

◄ **0453**
The sculptural design of the suspended
slat facade draws on the notion
of landscape in the city, a quality
visible in the graduated courtyard
garden and the building's silhouette
and layout. Plans for the ground
floor facing the street also include
a number of commercial spaces.

0454 ➤
The units' varying floor plans
and layouts indicate a number of
housing options; condominiums are
organized into townhouses with
private gardens, classic apartments or
penthouses with a spectacular view
of the old Friedrichstadt, Germany.

◄ 0455
Spacious, breezy transitions to the outside create an open residential experience in the middle of the city that, thanks to the variable heights of the different building levels, also offers an interesting succession of rooms.

◄ 0456
The generously sized apartments will face southwest, opening themselves to a view of the calm, carefully designed courtyard garden.

0457 ▼
Located on a consolidated number of smaller lots in the historical center of Jena, Germany, the separate structures allow for free access through the grounds.

◄ 0458
J. Mayer H. architects' design for the building, which will soon neighbor both Museum Island and Friedrichstrasse, reinterprets the classic Berliner Wohnhaus with its multi-unit structure and green interior courtyard.

0459 ►
Their placement on the outer edges of the plot defines a small-scale outdoor space congruent with the medieval city structure. Its outdoor facilities continue the building's overall design concept past the edges of the lot.

0460 ▲
The planned incorporation of commerce, residence and office enables a small-sectioned and flexible pattern of use that also conceptually integrates itself into the surroundings.

151

Johnsen Schmaling Architects

1699 North Astor Street
Milwaukee, Wisconsin, 53202, USA
Tel.: +1 414 287 9000
www.johnsenschmaling.com

◄ 0461
Smaller is better
Minimizing the footprint of a building limits the area of site disturbance as well as impermeable surfaces. For the Stacked Cabin, we organized the program components vertically, reducing the footprint by 50% over the horizontal configuration found in traditional cabin compounds.

0462 ▲
Reinvent and reuse
Reusing the bones of an old or obsolete building saves resources, embodied energy, and cost. The Ferrous House is the reinvention of a dilapidated suburban house; leaving the existing foundations and most of the perimeter walls intact, the inside was entirely reorganized to accommodate a contemporary lifestyle.

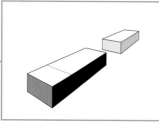

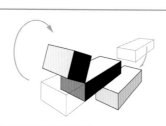

0463
Waste no space!
In order to minimize the size of the building, use every square inch intelligently. Built-ins may help utilize little corners that would otherwise be wasted space. In the OS House, we designed a built-in desk at the top of the stairs, offering a usable work surface in an area that would normally be impossible to use efficiently.

0464
Let it rain!
Instead of gutters and downspouts, which are usually connected to a municipal sewer system, use rain chains – they are easy to maintain and direct storm water from roofs to a rain garden or gravel bed, where it can percolate into the ground and replenish groundwater resources.

0465
Good insulation pays for itself
Use formaldehyde-free expanding foam insulation, which creates an airtight envelope with superior insulation characteristics. In the OS House, we used foam insulation made from agricultural by-products, thus eliminating the need for petroleum-based insulation products.

0466
Green roofs
Green roofs reduce storm water runoff by retaining it instead of feeding it directly into storm sewers; they minimize urban heat islands and increase the insulation properties of a roof. Using sedums instead of simple grass creates durable and low-maintenance plant covers that can survive even the harshest climates.

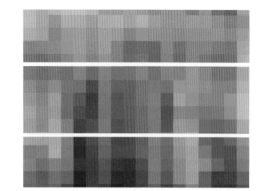
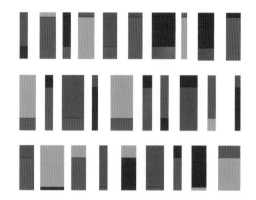

0467 ➤

Camouflage

Visual pollution is as destructive as that caused by fossil fuels. When designing a building in the countryside, allow it to blend in with its natural setting. At the Camouflage House, we pixelated images of the surrounding forest and transformed them into a panelized facade system that echoes the hues of the deciduous trees.

0468 ➤

Use recycled and recyclable materials

It is important to consider the origin of materials as well as their possible afterlife. Therefore, building materials should be recyclable and use recycled resources. We used steel panels for the Composer's Studio. The panels contain 80% recycled steel; at the end of the building's life cycle, the material will be fully recyclable into other steel products.

◄ 0469
Natural light
Maximize the amount of natural light that penetrates the building in order to minimize the use of electricity. North-facing clerestories with translucent glazing, like the one used in the Ferrous House, flood the interior with an even level of natural light, eliminating the need for artificial illumination throughout most of the day.

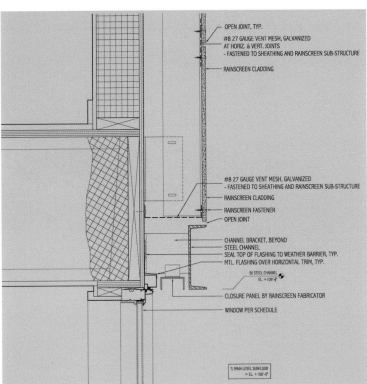

◄ 0470
Rainscreen facades
Rainscreen facades are sophisticated alternatives to single-layer facade systems, offering superior thermal performances and thus reducing heating and cooling costs. The rainscreen's open-air cavity between the insulated wall and the exterior cladding facilitates constant air movement between the two layers, eliminating the potential for trapped moisture or overheating.

Kendle Design Collaborative

6115 North Cattletrack
Scottsdale, Arizona, 85250, USA
Tel.: +1 480 951 8558
www.kendledesign.com

0471 ➤

Earth

Earth, as in the soil used for rammed-earth or adobe block construction, is possibly the most readily available, most indigenous and most sustainable material available. It provides thermal mass for storing heat during cooler months and deters heat-gain during warmer months. And it roots a structure to its site like no other.

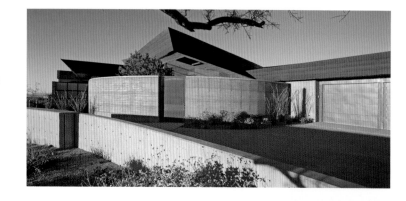

0472 ▼

Wind

Consider strategies that allow a home to "breathe." Placing window and door openings at opposite corners across interior spaces to enhance cross-air movement provides for passive cooling. Place operable windows low and high in order to facilitate a chimney effect for evacuating warmer air during the hotter months.

◀ 0473

Rain

Rain is something to be celebrated and cherished in the desert. In this example, Kendle uses folded roof planes in dramatic form to collect rather than shed rainwater, allowing its targeted distribution to surrounding desert plants

0474 ▶

Fire

Fire should not be used merely for its decorative beauty but placed to foster the use of indoor and outdoor spaces that are un-assisted by mechanical climate control. Used within a thermal mass, such as rammed earth, its warmth can be retained and distributed to surrounding living spaces.

◀ 0475

Water

Water is a scarce and precious resource in the desert. Placing it where it not only adds visual beauty but establishes a micro-climate of cooled and humidified air can enhance both indoor and outdoor comfort. Here, Kendle has positioned the pool to cool the air before it is drawn throughout the interior of the home, turning an amenity into a necessity, a demonstration of passive cooling.

◄ 0476
Shade can be a desert-dwellers best friend. However, properly designed indirect day lighting reduces the need for electric light, lowering energy consumption and bathing interior spaces in an ever-changing tapestry of light.

0477 ▼
The secret garden
Orienting interior spaces around a garden not only enhances the bond between man and nature, but it allows those spaces to benefit from the micro-climate established through retained moisture within the planted area. Here, Kendle uses a semi-enclosed garden to cool air falling from the adjacent mountainside before it is drawn through the interior spaces of the home.

0478 ➤
The outdoor room
Outdoor living areas designed as an extension of the interior volume expand the sense of spaciousness, blurring the line between interior and exterior spaces. This often permits reducing the interior conditioned volumes and expanding the use of non-mechanically climate-controlled living.

0479 ➤

Natural materials

Left in their natural state, natural materials create a timeless and sustainable pallet. Concrete, rammed earth, copper and sustainably forested or reused wood are best expressed without secondary finishes such as stucco or paint and require little maintenance, reducing the use of chemical cleaners.

◄ **0480**

Reuse, repurpose

Always consider reusing existing structures in a new way before discarding them for new. Here, a defunct restaurant is remodeled and repurposed as a spa and addition to an existing health club.

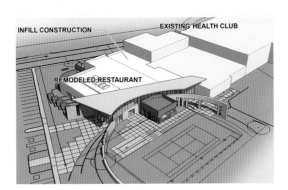

INFILL CONSTRUCTION
EXISTING HEALTH CLUB
REMODELED RESTAURANT

Kirkland Fraser Moor

Hope House, 1 Stocks Barns, Stocks Road
Aldbury, Hertfordshire, HP23 5RX, UK
Tel.: +44 1442 851933
www.k-f-m.com

0481 ➤
Beaufort Court is considered to be the first historical building to be refurbished for commercial purposes without any pollutants.

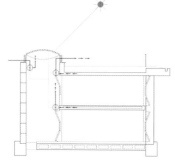

Ventilation section

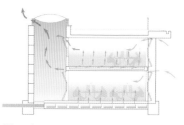

PV section

0482 ▲
The orientation of the complex, characterized by neutral carbon emissions along with other passive strategies, were taken into account when developing the project.

◄ **0483**
It is a historic example of the integration of renewable energies generated from the systems located on the site complex.

0484 ➤
The project includes a solar panel array, a biomass boiler, a heat store, a borehole with natural cooling and a 165-sq.-foot (15 m^2) wind turbine.

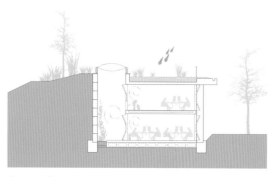

Green section

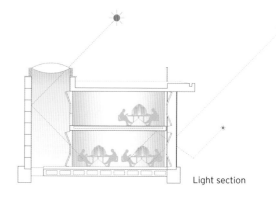

Light section

◄ 0485
The town of Fort William lies in the most strategic point of the West Highlands of Scotland.

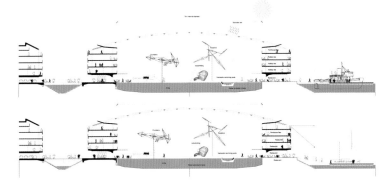

0486 ▲
It has a striking circular shape, enabling a micro-climate to be created to allow it to be used all year round, despite the low temperatures outdoors.

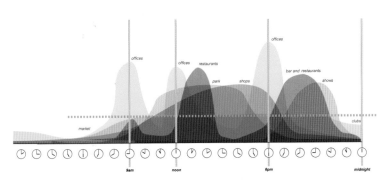

◄ 0487
The intersection of routes to the isles, the proximity to Glencoe and the tidal waterfront creates a leisure opportunity unmatched in Britain.

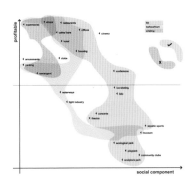

0488 ►
Graphics about culture versus profit.

0489 ▲
Land is reclaimed from the substantial tidal flats along the water's edge. The power consumed by this compound is obtained from the tides.

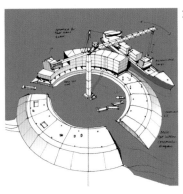

◄ 0490
The project includes plans to build a hotel, offices, shops, cultural centers and housing.

Living Homes

2910 Licoln Boulevard
Santa Monica, California, 90405, USA
Tel.: +1 310 581 8500
www.livinghomes.net

0491 ➤
To reduce water usage in your home, use low-flow fixtures in your kitchen and bathroom. Installing a gray-water storage and recycling system reclaims water from sinks, showers, and washing machines for irrigation in the surrounding landscape.

0492 ▼
Use Energy Star-rated appliances to reduce plug loads.

0493 ▼
Choose materials that are reused from existing projects, are biodegradable, contain recycled content, and/or are available locally.

0494 ➤
Use alternative materials for insulation, such as cotton denim, which does not contain harmful chemicals or irritants.

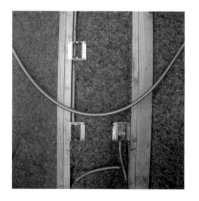

0495

Traditional paints contain VOC's (volatile organic compounds) that can cause irritation, nausea and even cancer. Low- or no-VOC paints and stains minimize off-gassing.

0496

If using wood, seek Forest Stewardship Council (FSC) certified wood. FSC is a non-profit that certifies forestries that practice responsible harvesting practices.

0497

Low-emittance (low-E) insulated glass doors reduce heat loss and heating costs, keeping you warmer in the winter and cooler in summer.

0498

Radiant-heating floors are a more efficient way to heat, and they do not release contaminants into the air the way conventional forced-air systems do.

0499

Create clean energy from the sun by using photovoltaics.

0500

Employ modular construction to reduce construction waste.

LMN

801 Second Avenue, Suite 501
Seattle, Washington, 98104, USA
Tel.: +1 206 682 3460
www.lmnarchitects.com

0501 ▽

Building elements can be used to provide protection from the elements through natural shading, insulation, and water retention. The City College of San Francisco Performing Arts Center's roof incorporates a deep overhang to shade the transparent public lobby while also providing 30,000 sq. feet (2,787 m²) of living roof that mitigates the solar exposure, reduces storm water runoff, and provides a habitat for local wildlife.

0502 ➤

Unwanted solar gain and discomfort can be avoided by careful siting and use of existing trees to provide shading. PACCAR Hall was intentionally located on a campus infill site in order to define public spaces and routes while maximizing the use of mature trees. Trees that could not be saved were reused within the project as part of featured furniture pieces.

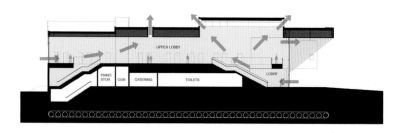

><

◄ 0503

Glazing systems should be designed to prevent unwanted solar heat gain. Marion O. McCaw Hall – home to Seattle's opera and ballet – incorporates a triple-height glass wall that connects the lobby to the exterior promenade. The high-performance double-layer wall system uses automatic blinds and vents to remove excess solar heat during the day and provide stunning drama at night.

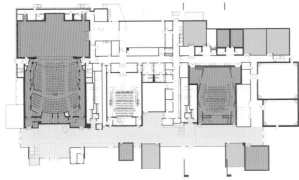

◄ 0504

Building elements can serve multiple functions. The City College of San Francisco Performing Arts Center's internal spine acts as the lobby for three performance halls; a venue for informal performances; a common circulation space to encourage mixing of faculty, students, and audience members; and as part of the natural ventilation strategy.

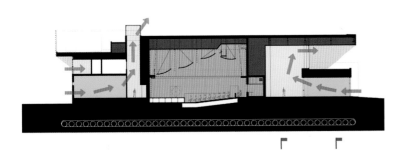

◄ 0505

Complex buildings require a complex mix of mechanical systems for optimum performance. This hybrid strategy is implemented in the LEED Gold-targeted City College of San Francisco Performing Arts Center, which uses natural ventilation, radiant ceiling panels, radiant floors and displacement ventilation to condition a wide variety of rooms, ranging from classrooms to large auditoriums.

0506 ▶

Buildings should improve their environment. Vancouver Convention Centre West is built on the site of an extensively contaminated industrial and rail yard and is 40% over water. In addition to remediation of the soil, the project's foundation creates a 1,500-foot (457 m) long artificial reef built in consultation with marine biologists to restore the natural shoreline and enhance the marine ecosystem.

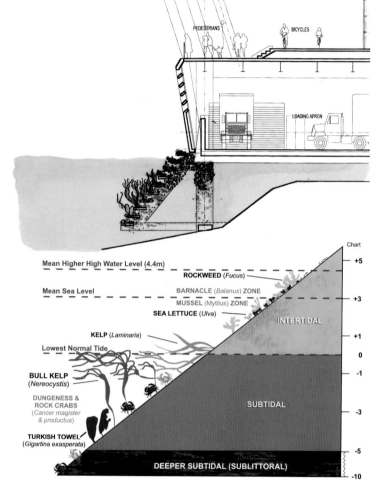

◀ 0507

Modern buildings can bring greenery back into our cities. The LEED Platinum Vancouver Convention Centre West incorporates an iconic 6-acre (2.4 ha) living roof that extends Vancouver's waterfront park and provides a low-maintenance habitat for birds and insects, including a colony of 60,000 bees, which provide honey for the center. The roof is watered with an ultra-efficient treated black-water irrigation system.

0508 ➤
Materials should be local and renewable. Vancouver Convention Centre West makes extensive use of locally harvested timber, including 165,000 linear feet (5,029 m) of Douglas fir Glulam beams and over 2-acres (0.8 ha) of hemlock cladding, which evokes the Vancouver waterfront's timber industry roots and the region's natural resources.

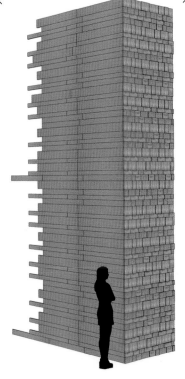

0509 ➤
Buildings should be designed as places for people. Vancouver Convention Centre West includes 400,000 square feet (37,161 m²) of walkways, bikeways, and plazas; 95,000 square feet (8,826 m²) of retail space; and is Vancouver's first major civic gathering space at the water's edge. The building allows users and casual visitors to enjoy outstanding views and connect with the natural setting of Coal Harbor.

0510 ⬈
Internal atria can dramatically reduce the area of external envelope for a building while providing significant natural daylighting. Careful skylight design can create a large impact with minimal amounts of glazing. These strategies are implemented at PACCAR Hall, the new LEED Gold home of the University of Washington's Foster School of Business.

Luis de Garrido

114 Blasco Ibáñez, 7-9
Valencia, 46022, Spain
Tel. : +34 93 356 70 70
www.luisdegarrido.com

0511 ➤
**Beardon Eco-House,
Torrelodones, Madrid, Spain**
It is 100% sustainable and self-
sufficient and, powered by solar and
geothermal energy. The house looks
like a plant sculpture, as its entire
exterior skin is green (walls, roofs and
sloped roofs). A plant wall-curtain has
been used for the first time, capable
of integrating vegetation with other
faces and glass on the facades.

0512 ▲
Green Box, Barcelona, Spain
Considered as the symbol of
sustainable architecture and the
best global reference. The house
appears as an extension of the
ground, as its sloping roof garden
has a very slight gradient. It is a
natural extension of the ground
and invites you to go for a stroll.

◀ **0513**
Green Box, Barcelona, Spain
This home is 100% sustainable and
bioclimatic with zero consumption
and zero emissions. It is self-sufficient
in terms of water. It has an infinite
life cycle, as all components are
assembled together easily, so that
they can be recovered, repaired and
used, as many times as necessary.

Winter

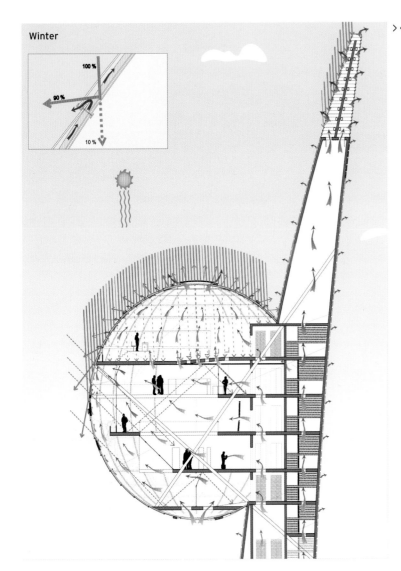

◄ 0514
**Berimbau, telecommunications
tower for the Olympic Games
of Rio de Janeiro, Brazil**
It is inspired by the berimbua, an
ancient Brazilian musical instrument,
and the capoeira, a Brazilian
dance, as it is designed to become
a reference in Brazil and move all
Brazilians. The building is 100%
sustainable and bioclimatic with zero
consumption and zero emissions.

0515 ▼
**Geoda 2055, self-sufficient
city, Mondragón, Spain**
The city is self-sufficient in terms of
water, energy and food. The most
characteristic feature is that the
water cycle has been expressed in a
sculptural form, creating a cascade
of water in which the buildings float
as if they were crystal geodes.

0516

Green2House, Shoeburness, United Kingdom

This is a sustainable home for a prestigious English writer. It is self-sufficient in terms of water and energy, with zero energy consumption and a high level of bioclimatic treatment. It is shaped like an open book, in tribute to the owner, and its innovative architectural style allows a high level of bioclimatic treatment.

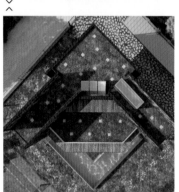

0517

Eye of Horus Eco-House, Sedir Island, Turkey

It is a house constructed for Naomi Campbell on one of the most beautiful islands in the world: Sedir. The home is bioclimatic, with zero energy consumption, and zero emissions and it is self-sufficient in terms of energy, water and food. Finally, all of its components are prefabricated and movable.

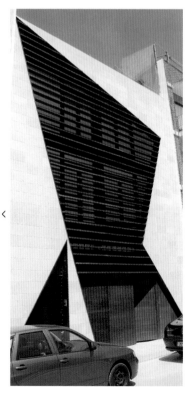

0518 ▲
Lliri Blau ecological housing development, Massalfassar, Valencia, Spain
It is Spain's first ecological housing development. It contains 129 low-price, bioclimatic and ecological houses. In its construction, only environmentally friendly and healthy materials have been used. The whole design is very unique and offers an important bioclimatic level, so that the houses are self-heating.

◄ **0519**
VitroHouse, Barcelona, Spain
The only house in the world built exclusively with flat glass. The housing is intended to illustrate the fact that true sustainable construction should be based on standardized and prefabricated industrial elements, assembled together.

Winter

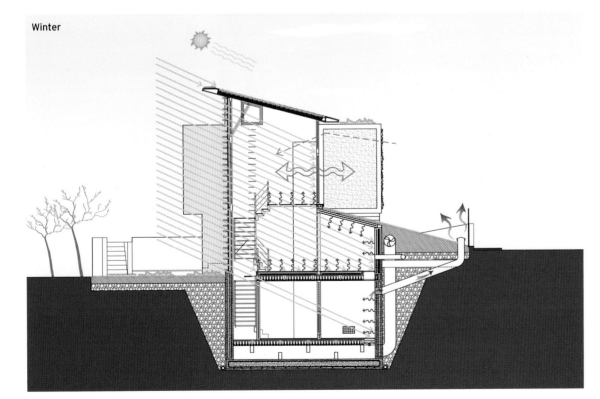

◄ **0520**
Unique design, high bioclimatic level
One of the most important features of the work by Luis De Garrido is that all buildings have a perfect bioclimatic performance due to their special design and care. Luis De Garrido has been laying the conceptual foundations of a new paradigm in architecture for over 20 years.

Manasaram Architects/
Neelam Manjunath

No 7, Aditigreenscapes
Survey No. 20, Venkateshpura, Sampigehalli Road
Bangalore, Karnataka, 560064,
India Tel.: +91 80 22792658
www.manasaramarchitects.com

0521 ➤

Rejuvenate the site
The project's site was a filled lake bed. Though not included in the project brief, we proposed the buildings along with a comprehensive lake rejuvenation scheme. Educate your clients on environmental issues with easy solutions; they will definitely accept it.

◄ 0522
Always feel a material before using it in your buildings. Listen to it. Then you can make it work the way you want.

◄ 0523
Try applying Indigenous systems to modern structures. It involves simple technology and high science and gives economic, sustainable solutions.

◄ 0524
Any project design should start with an in-depth analysis with several system maps that show the relationships. All these maps should lead to some conclusions, and that should give you the design concept or the starting point of the design solution.

0526 ▲
Local materials and building technologies are easy to source, easy to work with, weather well in their climate and are economical; hence, they produce effective and sustainable solutions.

0525 ▼
Stone, mud, bamboo, wood, and other such natural materials are going to be the main building materials of the future. With the rise in urbanization, waste or recycled waste could be the future, with large building and infrastructure projects being undertaken in cities.

◄ 0527
Understand who your real clients are. Center your design on them.

0528 ►
Building a glass box, or such an unsustainable building, and then installing mechanical systems, such as air conditioning, for comfort, light, etc., then making the mechanical systems energy efficient does not lead to a zero carbon building. Try making your building zero carbon and not dependent on mechanical systems. That is sustainable architecture design.

0529 ►
Each geographical region in this world has suitable building materials and it's own particular construction system. It is nature's way of giving solutions to us. Look for it, listen to it.

0530 ►
Sustainable development is local solutions to global problems and not global solutions to local problems. These global solutions are always energy intensive, uneconomical, difficult to execute and hence unsustainable.

Mario Cucinella Architects srl.

3/A Via Barozzi
Bologna, 40126, Italy
Tel.: +39 51 6313381
www. mcarchitects.it

0531 ➤

The SIEEB building is a platform for the bilateral cooperation between Italy and China in the fields of energy and environmental preservation efforts. It is also a showcase building for reducing CO_2 emissions in the construction sector in China.

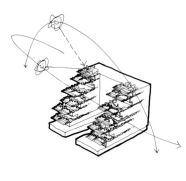

◀ **0532**

The design incorporates passive and active strategies in its form and skin to control external environment and improve interior air quality.

◀ **0534**

The offices and laboratories located on the upper floors have terraced gardens that also use cisterns. Together with overhanging structures, where the photovoltaic panels are located, they act as sunshades for the floor directly below.

0533 ▼

The building features a bioclimatic design. It combines the use of glass for natural lighting with renewable energies and rainwater collection by means of roof cisterns.

0535 ➤

The north-facing facade is closed and highly insulated, given that the cold winter winds come from this direction. The facade becomes more transparent toward the south. On the east and west sides, sunlight and radiation are controlled through the double-glass skin.

0536 ➤
These sustainability efforts enable the SIEEB to emit around 550 short tons (499 metric t), of CO_2 per year, compared to the average of 30,860 tons (27,996 metric t) emitted by a conventional building, with a 50% saving on air-conditioning usage and costs.

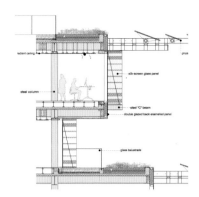

◀ **0537**
This residential building was designed for the 2008 exhibition Habiter Ecologique held at the Cité de l'Architecture in Paris, France. The building is 19,160 sq. feet (1,780 m²) over six floors and is located in Boulogne Bilancourt, just outside Paris.

0538 ➤
Shared services are on the ground floor, such as the kindergarten, restaurant and bar, and on the roof there is a garden shared by all the inhabitants.

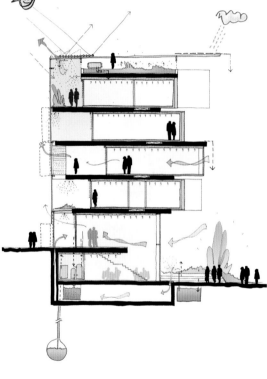

0539 ⬆
The environmental strategies employed, such as photovoltaics, certain materials and insulation, have allowed a building that is energy positive, producing 0.6 kW/sq. foot (6 kW/m²) per year more than it consumes.

0540 ➤
It is conceived as a series of overlapping timber layers. One facade has protruding "boxes" that create a dynamic elevation with private protected terraces. The opposite side is calm and ordered. The spaces flow into one another and link the two sides of the building.

Mark Kingsley Architects

Piccadilly House, London Road
Bath, BA1 6PL, UK
Tel.: +44 teh 1225 326442
www.kingsleyarchitects.com

0541 ➤
Environmentally responsive architecture needs to work in synergy with the context of its site. Form and the arrangement of function respond directly from a topographical understanding of site and its available natural resources.

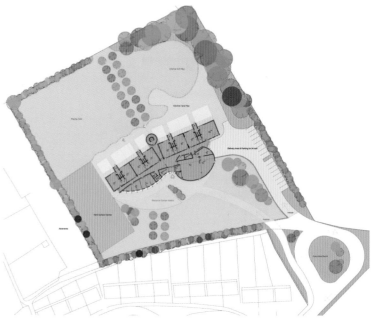

◄ **0542**
Obtain a thorough understanding of the macro- and micro-climate issues associated with a site. Utilize the natural environmental strengths of a site to inform the design. The resultant architecture should then reflect and work in harmony with these strengths.

◄ 0543
Urban environments offer the opportunity to insert environmentally responsive architecture that is in keeping with the architectural grain of the local town or cityscape. Reduce the urban heat island effect and improve biodiversity by designing in green roof and wall systems.

0544 ➤
Sustainable architecture requires appropriate levels of financial investment. Additional costs, beyond the basic building costs, have an impact on time, quality and durability. This investment will then result in financial savings for the residents during the life of the building (typically 50 or more years) along with other brand, community, well-being and environmental benefits.

0545 ▼

High levels of natural daylight should be designed in and positioned to prevent overheating from summertime solar radiation. Stacked ventilation should also be included to allow a building or dwelling to stay naturally cool, without mechanical means, in its summertime mode.

◄ 0546

Utilize space, transparency and views in designs to enhance lives. Specify and build with as many locally sourced natural (non-synthetic) materials as possible ensuring they are renewable, durable, of low environmental impact and recyclable.

0547 ➤

In northern hemisphere countries, adopt a fabric-first design approach. A highly thermally efficient and airtight external fabric will require very little energy to heat. A low-energy ventilation strategy should then be adopted; the type will depend on the use and an occupant's thermal comfort requirements.

◄ 0548
Challenge conventional design norms to reduce the resultant environmental impact of architecture and environments. Question perceived requirements and how they could be rethought requiring less or no energy.

0549 ▼
Existing and period architecture can be successfully refurbished or retrofitted to enhance user comfort and reduce its environmental impact without needing to be demolished.

0550 ▲
Question how, why and where we live. Is it for socio-economic reasons we live the way we do? Why is the design of our habitats and dwellings so constrained by society? How could we live in a manner that is more in harmony with nature?

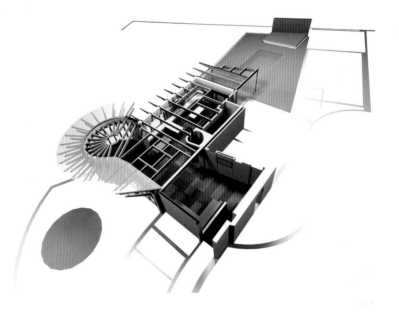

Mark Merer

Welham Studios
Charlton Mackrell
TA11 7AJ Somerton, Somerset, UK
Tel.: +44 1458 223341
www.markmerer.com

0551 ➤
Climate, orientation, position on the planet from a geographical and cultural point of view should be the starting point of any design. It should embody the spirit of a place and time.

0552 ➤
How we place a building in the context of a site is one of the most important aspects of the design process and is a reflection of our attitudes. Look for your own relationship to the surroundings.

0553 ▲
Gardens need very little lighting, so make use of what is available. To see the night sky requires very low levels of illumination, and the night sky can be far more evocative than artificial light.

0554 ▼
Light is one of the most important aspects of a design, as it informs us of the space. It can provide passive heat while keeping us connected to the outside.

◄ **0555**
A truly sustainable building is not just about energy consumption and responsible materials. The design itself has to aim at an aesthetic quality to ensure its existence in the future.

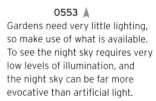

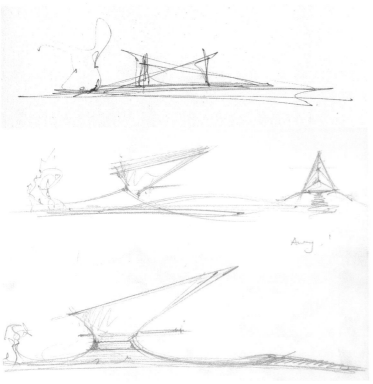

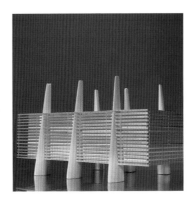

◄ **0556**
Always materialize the idea in model format as soon as possible.

◄ **0557**
Place faith in the design process and use drawing as an essential tool, not for presentation but for exploring the ideas.

0558 ⋀
Nature has many examples of the union of object and environment, which can often help us find solutions to how we build within different contexts.

0559 ➤
I see no boundaries between any three-dimensional construct and spend my time observing land use, construction and development. I find the three dimensions universal, a conduit that enables us to see the physical world in more than scientific truths.

0560 ⋀
Local culture and traditions are vital considerations. When using traditional formats, be careful not to just supply clichés. That will only make a mockery of the heritage.

Max Pritchard Architect

ABN 91 090 808 815
PO Box 808
Glenelg, South Australia, 5045, Australia
Tel.: +61 8 8376 2314
www.maxpritchardarchitect.com.au

0561 ➤
Preserve and respect the local environment
A 656-foot (200 m) long structure of 21 luxury resort suites steps down to follow the slope of the land with minimal land and vegetation disturbance. A continuous ramp connects the suites.

0562 ▼
Collect and use rainwater
Over 264,170 gallons (1 million L) of water is stored from roof collection for use in this luxury tourist lodge, which reduces dependence on central services.

0563 ➤

Minimize disturbance of the land
Use a "touch the earth lightly" approach. Here, the house is supported from the ground at only four points. Construction can be largely prefabricated off-site.

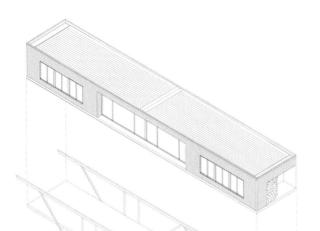

◄ 0564

Use local materials
Stone is collected on the site and incorporated into the building to form walls, fireplaces, etc. The timber framed structure is sourced from local plantation pine.

◄ 0565
**Maximize sun access
for winter heating**
Orientating all living areas to the
north (in the southern hemisphere)
allows the low winter sun to penetrate
inside and provide most of the winter
heating. The daytime heat is absorbed
into the dark concrete floor and
re-radiated at night. Well-insulated
walls and windows help maintain
comfortable conditions naturally.

0566 ➤
Reuse existing structure and materials
A 26-foot (8 m) diameter and 20-foot
(6 m) deep concrete rainwater tank was
constructed 60 years ago to supply
the city's water. It has been adapted
by building another level on top and
cutting openings in the tank wall to
form an economical family home.

0567 ▼
**Use vegetation for economical
but attractive summer shade**
Deciduous vines climb over a
light structure, providing shade
in summer but allowing winter
sun to penetrate indoors.

0568 ▲
Connect the inside with the outside
In many climates, outdoors is often the most pleasant environment, as long as there is protection from the sun, wind and rain. Even in dense cities, courtyards can provide a natural oasis. Give people a chance to open a door or window before they reach for the air-conditioning remote.

◄ 0569
Design for the climate
Sited on a hill with expansive sea views, the house is buffeted by winds from many directions. Creating multiple outdoor spaces means there is always a sheltered outdoor option.

0570 ▼
Plan efficiently
A small house designed as a three-level tower fits among existing trees with minimal land disturbance. Design multiple-use spaces and maximize views. Sliding doors and large windows allow relatively small spaces to appear expansive and encourage appreciation of the outdoors.

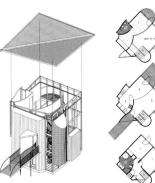

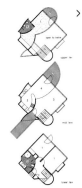

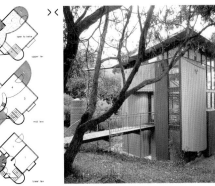

Leddy Maytum Stacy Architects

677 Harrison Street
San Francisco, California 94107, USA
Tel.: +1 415 495 1700 ext. 302
www.lmsarch.com

0571 ➤
Pushing to net zero energy
Vai Avenue Case Study,
Cupertino, California

0572 ▲
Transforming historic resources
Bay School, The Presidio of San
Francisco, California

◄ 0573
Adaptive reuse/harvesting built resources
California College of the Arts, San
Francisco, California

0574 ▼
Buildings that teach
Nueva Hillside Learning Complex,
Hillsborough, California

◄ 0575
Fostering biodiversity/rebuilding Native Habitats
California Shakespeare Theater, Orinda, California

◄ 0576
Preserve history and place/ sustainable transformation
Cavallo Point, Lodge at the Golden Gate, Sausalito, California

0577 ▲
Synergy between universal design and sustainable design
Ed Roberts Campus, Berkeley, California

0578 ▼
Connecting to the rhythms of the day and the seasons
Homer Center For Science & Student Life, Atherton, California

0579 ▲
Social sustainability: Building a just and sustainable future for all
Plaza Apartments, San Francisco, California

0580 ▼
Biophilic design: Making abstract connections to nature's forms
Climateworks, San Francisco, California

Michael Jantzen

www.humanshelter.org
mjantzen@me.com

◄ 0581
The project will be approximately
150 feet (46 m) high, 250 feet
(76 m) long and 120 feet (36.5 m) wide.

0582 ➤
The pavilion is designed to celebrate
the return of the sun's rays every
day to the surface of the earth.

0583 ▼
The structure will consist of 12 precast
concrete rectangular columns,
symbolizing the rays of the sun.

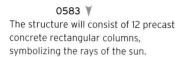

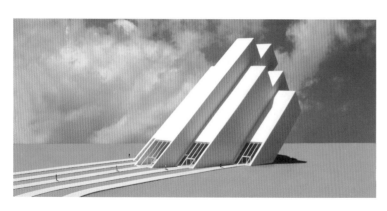

0584
The design of this winery explores ways to integrate alternative energy collection and storage systems into commercial architecture.

0585
The pavilion should be erected on a site in a temperate climate, and could function as a winery, wine-tasting center or a venue for special events.

0586
The steel arches and horizontal supports would be covered with glass panels that open and close automatically, coated with photovoltaic material to control natural ventilation.

0587
The Eco-Tower is a conceptual design for a public meeting place with a special Wi-Fi network that might be located on a university campus or in a public park.

0588
The structure can be constructed using prefabricated glass and steel.

0590
The tower would have seven platforms, six of which would be accessed via a spiral staircase in the middle. Finally, on top of the roof there would be a wind turbine.

0589
The observation structure would have an approximate height of 120 feet (36.5 m) and could be built of concrete and steel.

Minarc

2324 Michigan Ave
Santa Monica, California 90404, USA
Tel.: +1 310 998 8899
www.minarc.com
www.mnmMOD.com

0591 ➤

Enjoy life
Spend family time together at home.
Playing croquet is both healthy for your
mind and body and very sustainable.

◄ 0592

Layering
Try to avoid layering. No additional
layering of decorative materials,
like paint, tile and carpets are
needed and only add unnecessary
chemicals and waste.

0594 ➤

Simplicity – simple is beautiful
Easily deconstructed and recycled parts at the end of building's life cycle create zero waste, diverting all materials from the landfill. Design a very energy-efficient building envelope by selecting an external wall system and door/window package with high R- and U-value.

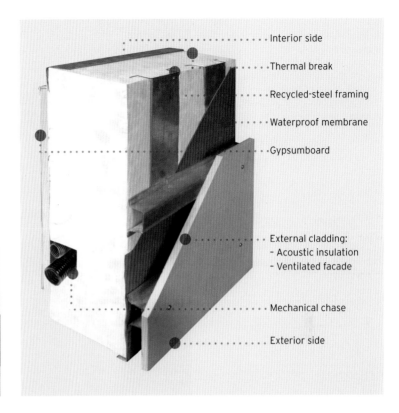

- Interior side
- Thermal break
- Recycled-steel framing
- Waterproof membrane
- Gypsumboard
- External cladding:
  - Acoustic insulation
  - Ventilated facade
- Mechanical chase
- Exterior side

0593 ▼

Recycle
Giving something a second chance is good for the soul. Try it sometime!

> <

0595 ➤

Use what you have
Design environmentally and economically sensitive structures: "If a building's orientation, massing, window area/shading, insulation arrangement, and air tightness are not properly optimized, no amount of mechanical system makes the building a 'low-energy' building. And if it is not a low-energy building, it is not a sustainable building."

0596 ➤

**Be responsible – respondable
(ability to respond)**
Sustainable building is a social
responsibility to reduce the building
carbon footprint. Select green
materials, systems and technologies
that are sustainable.

0597 ▼

Functional
Chairs that have their own space in
the island. Time is valuable, so our
environment should be functional
and, therefore, sustainable. This saves
time, effort and frustration, and it
gives us more time with family.

0598 ➤
Change is good
Use alternative construction methods, such as thermo mass, which reduces the mechanical load, energy use and cost.

0599 ▼
Life style choice
You should not downgrade your lifestyle as an easy solution. For example, is there anyway you can live close to work. This saves time and energy. It is preferable to live where you can beat traffic on a bike. That would give you a workout and transportation at the same time. Can you put a water filter on your faucet to eliminate bottled water? Make your own sparkling water.

0600 ▲
Common sense
Passive and active design principles are practical and sensible.

MVRDV

10 Dunantstraat
Rotterdam, 3024 BC, The Netherlands
Postbus 63136
Rotterdam, 3002 JC, The Netherlands

Tel.: +31 10 477 28 60
www.mvrdv.nl

0601 ➤

**Spijkenisse public library,
The Netherlands**
The program consisted of building
a public space that incorporated
intelligent, low-energy environmental
technology while not requiring
an overly large investment.

◄ 0602

**Spijkenisse public library,
The Netherlands**
The building has a timber frame with
glass walls. The floor is fitted with
radiant floor cooling that features cold
water flows to offset the heat that
accumulates through sunlight and
body heat. This system operates by
releasing cold air under the shelves.

SUMMER

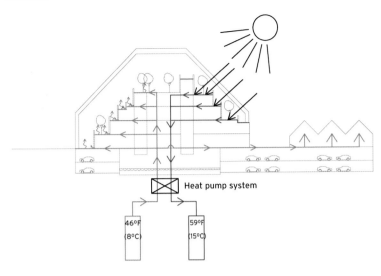

Heat pump system

46°F (8°C)   59°F (15°C)

WINTER

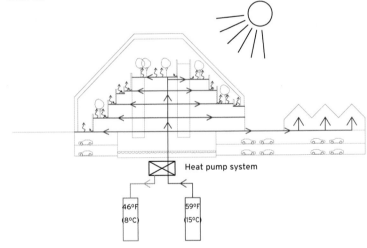

Heat pump system

46°F (8°C)   59°F (15°C)

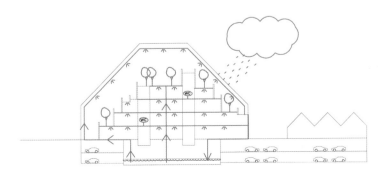

◄ 0603
**Spijkenisse public library,
The Netherlands**
The cool exterior is heated in summer by the adiabatic effect. The system sprays water into the interior air, cooling it. This cool air enters the heat exchanger and extracts heat from the incoming air. Additional cooling of incoming air in summer is provided by the cold accumulated in the underground collector.

0604 ▲
**Spijkenisse public library,
The Netherlands**
The facades have automatic sun protection systems and a ventilated second skin formed by sunshades. This system filters 90% of solar radiation, preventing excess heat. The roof collects rainwater, which is used in the toilets, to water plants, for fire protection and the adiabatic air-conditioning system.

0605 ➤
Montecorvo, Eco City, Logroño, Spain
In 2007, MVRDV together with the Spanish office of GRAS joined a competition for a sustainable urban extension to the city Logroño, a medium-size city of approximately 130,000 inhabitants, in the wine region of La Rioja in the north of Spain.

0606 ▼
Montecorvo, Eco City, Logroño, Spain
The 138-acre (56 ha) site, just north of Logroño on the two small hills of Montecorvo and la Fonsalada, has a lot of potential for a new neighborhood. Besides the fact that the hills provide a beautiful view over the city, the slopes are orientated to the south, making solar energy easy to generate.

0607 ▲
Montecorvo, Eco City, Logroño, Spain
The program consists of approximately 3,000 houses and a complementary program: schools, social buildings, sports facilities, all developed in a sustainable way. By producing all the energy needed on-site, the new neighborhood achieves a carbon-neutral footprint.

◄ 0608
**Gwang Gyo Power Centre
site, South Korea**
The Gwang Gyo Power Centre
site is situated 22 miles (35 km)
south of Seoul. It is part of a major
development for a new mixed city.
Two centers will be realized in this
development. One of these will be
the new power center; the other is
the new central business district.

0609 ▼
**Gwang Gyo Power Centre
grottoes, South Korea**
This shifting of the floors causes, as
a counter-effect, hollow cores that
form impressive empty grottoes
that are open at the bottom.
Here they form lobbies for the
housing and offices, plazas for the
shopping center and halls for the
museum and leisure functions.

◄ 0610
**Gwang Gyo Power Centre,
the Green Powercentre, South Korea**
Perhaps we can create a landscape
on top of the new program that
expands the green qualities and
links the surrounding parks. It would
thus turn the site into a park itself.
This park can form an aquatic and
energy buffer. It can hold the water
for cooling and cool the buildings,
reducing the energy demands.

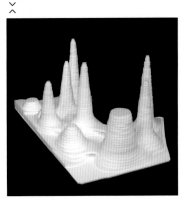

197

Nicholas Burns Associates

Tel.: +65 6738 0064
nb@nicholas-burns.com
http://www.nicholas-burns.com

0611 ➤

Every element should have multiple functions; a screen provides shade, creates privacy, allows cross-ventilation, and is powerful design element and light filter.

0612 ➤

Factor adaptability into possible future functions and allow for others that you haven't considered by making the building dynamic.

◄ 0613

Design using durable and recycled materials that evolve and improve over time. This increases the usable life span of the building without the need for renovation or repair.

0614

Passive is better than active
Use passive means first. Block
summer sun, allow in winter
sun, optimize thermal mass and
accommodate cross-ventilation.

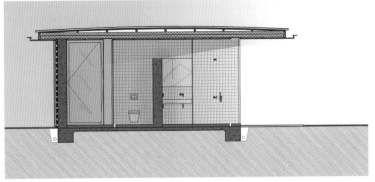

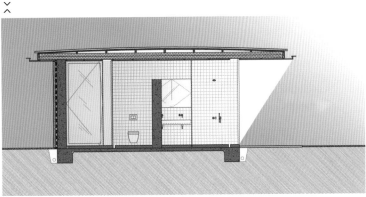

0615

Don't rely on insulation
Termal mass complements insulation
and stabilizes the internal climate.

0617 ➤
Reuse
Look for ways to improve the functionality and climate response of older buildings.

0616 ▼
Learn from the past
Look to indigenous cultures. The way these cultures have adapted to the landscape and climate is a great resource.

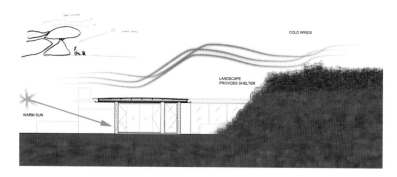

◄ 0618
Engineering
Avoid engineers that use a cut-and-paste methodology. Buildings are often over-engineered by up to 30%, increasing the embodied energy of the project unnecessarily. Find engineers that work and think hard, calculating structural loads for all elements of each individual project.

0619 ➤

Grow food

Small courtyards, roof space and even a balcony can provide meaningful quantities of produce, connecting you to seasonality, and increasing diversity and health while reducing your carbon footprint.

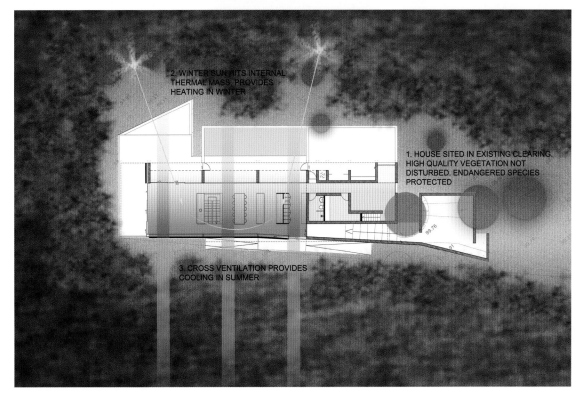

2. WINTER SUN HITS INTERNAL THERMAL MASS, PROVIDES HEATING IN WINTER

1. HOUSE SITED IN EXISTING CLEARING. HIGH QUALITY VEGETATION NOT DISTURBED. ENDANGERED SPECIES PROTECTED

3. CROSS VENTILATION PROVIDES COOLING IN SUMMER

◀ **0620**

Site

Start with a detailed site analysis and use the landscape to your advantage. Capture the sun and cooling breezes, and look for areas that offer the opportunity to build with minimal disturbance to flora and fauna.

OFIS Arhitekti

Ljubljana , 1000, Slovenia
Tel.: +386 1 4260084
www. ofis-a.si

0621 ➤
This cabin is located in a small
village that is part of the Triglav
National Park, Slovenia. There are
very strict building rules in this area,
particularly to prevent deterioration
of the natural landscape.

0622 ▼
The house, measuring 20 x 36 feet
(6 x 11 m) and with a gabled roof at
an angle of 42 degrees, has the same
dimensions and materials as the
structure that was originally on the site.

◀ **0623**
The materials, including stone and
timber, are locally sourced and used
in a way that is consistent with
the local style of architecture.

0624
The residence has a logical interior layout. The staircase wraps around the central fireplace, which provides heat for both floors.

0625
Urban growth and suburban expansion is causing the rural towns around Ljubjana to merge with the capital. This development of single-family residences maintain the typical layout of rural settlements, with small branching streets connected by paths.

0626
The temperature is regulated by the fireplace and passive systems, such as the windows being oriented toward the sun, and the placement of black foil behind the wood to absorb heat and conduct it to the interior.

0627 ▼
Rainwater is collected by means of vertical pipes inserted into the wooden posts.

0628
One of the first ideas was to use the orientation of the houses and their openings to create cross-ventilation.

0629
The materials are locally sourced, and the finishes are water-based and nontoxic. The water used for watering the green areas is harvested rainwater.

0630
Natural light was also maximized, particularly in winter, although the overhanging features protect the interior from excess exposure to the sun in the summer.

Office of Mobile Design by Jennifer Siegal

1725 Abbot Kinney Boulevard
Venice, California, 90291, USA
Tel.: +1 310 439 1129
www.designmobile.com

Collaborators: Taliesin
Design/Build Studio, M.P
Johnson Design Studio

0631 ➤
Ninety years ago, Frank Lloyd Wright introduced a pioneering plan to build houses with prefabricated parts: precut structural frames, mass-produced furnishings, etc. The First World War frustrated the project, and only a handful of houses were built.

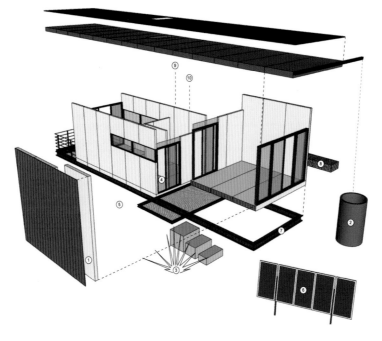

0632 ▼
The Taliesin Mod.Fab prototype is an attempt to follow in the path of this great architect and is seen as an example of a simple, elegant, and sustainable home.

0633 ▲
This home was built using structural insulated panels, which make the production and assembly processes fast and economical.

0634 ➤
It can also function off-grid, taking advantage of natural ventilation, photovoltaic panels in the yard, rainwater collection and gray water.

◄ 0635
All of these features demonstrate the feasibility of sustainable architecture, making good use of the resources offered by nature.

0636 ▼
The photovoltaic array in the yard provides electricity to meet the occupants' needs.

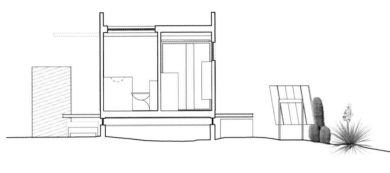

0637 ▼
Additionally, most of the materials used follow sustainability criteria. The paints are free of volatile organic compounds (VOCs) and the adhesives are nontoxic.

0638 ▼
The glass is low-emission, which prevents infrared radiation from passing through and heating the interior in summer. It also prevents heat loss from the interior in winter.

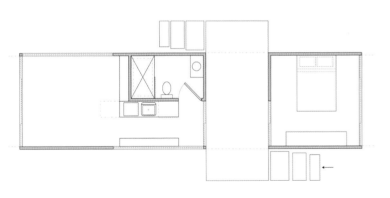

0639 ▲
Two of the facades are exposed, while the rest are more enclosed and private.

◄ 0640
The roof cantilever creates a porch area that offers shelter and shade.

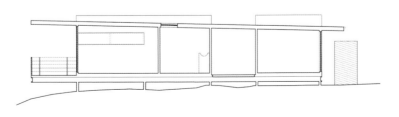

OLGGA architectes

14 Rue de l'Atlas
Paris, 75019, France
Tel.: +33 1 42 40 08 25
www.olgga.fr

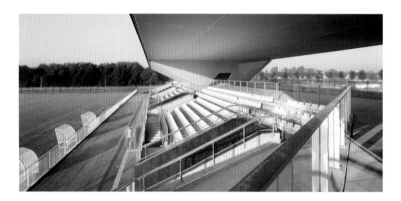

◀ 0641
The Stadium du Littoral, built at Grande Synthe, France, is the first facility north of Paris that brings together five sporting disciplines under one roof. This sports venue has stadium seating capacity for 617, two club houses, locker rooms, a multi-purpose hall, and administration offices.

0642 ▼
The project was designed as a single rectangular block in response to the unique program and thermal load specifications laid down by the clients.

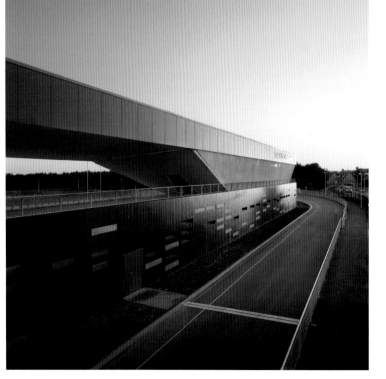

◀ 0643
Our sustainable approach combined with the clients' wishes regarding ecological impact, led to the creation of a building that is at the cutting edge of environmental friendliness. The Stadium du Littoral has very low energy consumption and achieves significant savings. The stadium makes use of durable materials and was built in compliance with French Très Haute Performance Energétique (THPE 2005) energy performance standards. Reinforced insulation, two condensation boilers, and a double-flow ventilation system to prevent heat loss. Low-energy lighting was installed in the interior.

Vertical section

1. Wall construction: Round log construction, halved, maritime pine, processed in an autoclave, diameter of 6 1/4 inches (160 mm).
   Sub-construction: Scots pine vertical lathing, 1 1/4-inch (30 mm) sealing.
   Solid five-ply larch wood panel glued crosswise, 4 inches (102 mm).
2. Roof seal
3. Laminated glass 1/16 x 1/6 inch (2 x 4.4 mm).
4. Scots pine glass strip 3/4 inch (20 mm).
5. Acacia cross-beam, 6/6 inches (150/150 mm).

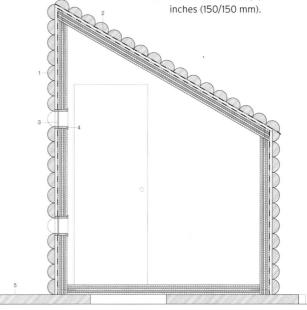

0645

Flake House, a nomadic dwelling that complies with standards for road transport, was designed to embellish the place where it is set up and to turn it into a singular scene. The house is split into two parts in order to establish a radical spatial boundary between living and functional areas while also creating an unexpected entrance. The interior finishes were purposefully pared down with a smooth appearance to contrast with the traditional look of the exterior log siding.

0646

Building program for 100 student homes in Le Havre, France, using metal shipping containers according to the clients' specifications.
Certification: French high energy performance THPE 2005 standard and EcoCampus certification.
Clean site, solar hot water, reinforced insulation, and rainwater capture.

0647 ➤

EvolutiV' is a design for a modular ecological house with a timber frame. The model is based on energy consumption of 4.5 kW/sq. foot (48 kW/m^2), per year. Energy performance is guaranteed by different systems:

- South facing outlook that enables solar radiation to heat the house in winter;
- A heat pump, combined with heat exchanger, provides an ideal ambient temperature;
- Solar panels on the roof produce sanitary hot water;
- A rainwater capture tank installed inside the wood pile feeds the toilets, washing machine and faucets for watering the garden;
- Wood wool insulation.

◀ **0648**

Our management and environmental goals led us to design a complex of 21 low energy homes in Beuvrages, France, that offer an alternative to the typical layout of suburban housing developments. We used an innovative construction system: timber frame with brick facing. This system takes advantage of the building and thermal qualities of wood in combination with the durability of brick. Low-energy certification: solar hot water, wood-burning stove for thermal inertia with gas condensing boiler; R-20 7 3/4-inch (20 cm) thick wood wool insulation; water-saving appliances.

0649 ⚠

Program for the construction of 100 residential units and commercial spaces in Caen, France.
Certification: Effinergie low-energy label and French H&E certification Level A; polystyrene panel external insulation on vertical walls with systematic correction of thermal bridges, installation of high-energy performance aluminum and wood window joinery.

> <

◀ **0650**

This business incubator in La Rochelle, France, specializes in renewable energy companies. The main facade is of solid wood slats that create a filter to protect the interior from the sun. These wood slats are stacked some distance away from the largely glass building front. This distance combined with the spacing between the slats enables solar incidence to be optimized in summer without obstructing solar incidence in winter. Low-energy certification: passive solar strategies, double-flow ventilation, geothermal gas heating, rainwater capture and green roof.

Paratelier

Leonardo Paiella, Monica Ravazzolo
8A Rua São Mamede
Lisbon, 1100-582, Portugal
Tel.: +351 218822075
www.paratelier.com

0651 ➤

Let nature be the source of inspiration and a starting point of a design. In the case of 'Guest House, it was the pine wood, the sand dunes of the Atlantic ocean and the local climate, that set the direction of design and conditioned the architecture.

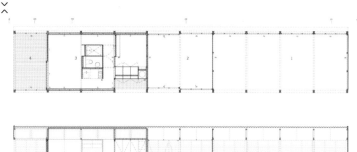

SIDE SECTION - S01

0652 ▼

To enhance the vivid colors of the surrounding nature, architecture is made transparent, semi-transparent and opaque, with the ability to transform itself depending on the needs and the current weather conditions.

◄ 0653

The architecture, entirely made of the local pine wood, is designed as a light and simple construction that, besides considering the Portuguese architectural culture, respects the natural environment in the most sensitive, aesthetic and creative way.

◄ 0654
Everything matters. Along with the wood, other organic materials like cork and eco-paint were used to create this highly adaptable, sustainable and recyclable architecture. Self-ventilating walls represent another way for architecture to interact with nature, paying respect to the importance of environmental sustainability.

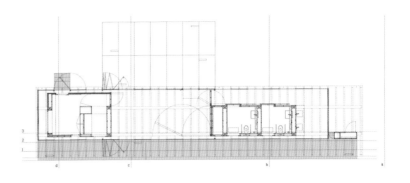

0655 ▲
In the case of the Pavilion of Almada, sustainability was also considered through the sociocultural context that is necessary to achieve holistic environmental balance.

0656 ▼
Flexible and foldable, the pavilion integrates itself into the park environment as a semi-transparent structure, where full is void and where inside is outside.

◄ 0657
The pavilion was exclusively realized in eco-compatible material. Only one material was used, wood. Outside, rough, untreated, rugged, plain and full wood was used. It is transformed into the surface and increasingly defined and refined. The bands are made from local pine, left raw to visually limit their environmental impact on the landscape.

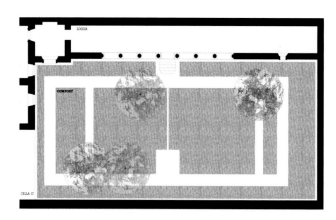

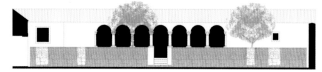

0658 ▲
Creatively demonstrate the natural processes that are crucial to the earth's functioning. The Ortus Artis project rises from the desire to subvert the common idea of "garden = green" and suggests new sensations and images by reinterpreting the issue.

0659 ▼
The core of the project is the use of different levels to create the sensation of walking through paths on soft earth that was just stirred. In reality, there was no real excavation; rather, the ground level was lifted into 6 1/2-foot (2 m) "walls" topped with a layer of compost.

◄ 0660
The brevity of human life as an integral part of the natural process is one of the keys of ecological thinking. In the center of the project area, still surrounded by the "walls" of compost earth, there is a basket full of apples and an invitation for the visitor to eat them and throw the cores in the compost walls.

Paterson Architects

6 Darnaway Street
Edinburgh, EH3 6BG, UK
Tel.: +44 131 220 1088
www.patersonarchitects.com

0661 ➤

Less Is More II

Why use a dozen materials when one will do? The simple act of building in timber retains CO_2 for the life of the building. Use screws, not nails, and the timber can be reused indefinitely. When forestry is managed, felled timber is replaced with new planting so balance is maintained indefinitely.

0662 ▼

More Is More I

Who can predict the future for a building? In a Victorian tenement, you might find: young or elderly single people, couples, families, an accountant's office, a dentist's office, an online retailer. The list is endless. In a newly built apartment block in Scotland you will find single people and couples. It's not just developers who like it that way, so do our city planners.

◄ **0663**

Loose fit

Our house plans are always flexible and allow for long-term adaptability. One of our key aims is to create lifetime homes.
We also incorporate space to recycle, space for bikes, space for things you don't need now, but might need later?

◄ **0664**

Less Is More I

By their longevity alone, existing buildings are already sustainable. They have stood the test of time and taste. Huge windows, large rooms, high ceilings, period detail and "character," can all be found, often on sites on which it would now be impossible to obtain permission to build. However, they often no longer perform or function as well as they might otherwise, either in layout or thermal performance. Reuse, refurbish and adapt.

0665
More Is a Bore I

Good, functional and appropriate design lasts a lifetime. Our aim is to always keep it simple. Our design solutions always derive from the site. In our project Three Seton Mains, the living accommodation is elevated for the view, as this diagram clearly shows. The resultant house is configured around this key design principle.

0666 ➤
More is a Bore IV

"Product," has a limited lifespan. Photovoltaics, wind turbines, water wheels, boilers, heat exchangers, etc. will fail or become obsolete. Invest in insulation and air-tightness; high-performance windows and durable, sustainable materials. Invest in good, appropriate, site-specific design.

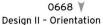

◄ 0667
Design III – Natural Light

All our projects seek to maximize the amount of natural light to the interior habitable spaces, something we believe is imperative in our part of the northern hemisphere. We undertake sun path diagrams to help decide on window locations. Clients have commented favorably about not having to put their lights on during Scottish daylight hours.

0668 ▼
Design II – Orientation

All our projects pay particular attention to their orientation. Our project, House on a Hill is a good example of this. Large south-facing windows light the principle spaces. East- and west-facing windows light the bedrooms for morning and evening sun. In the Scottish tradition, rooms on the northern elevation are ancillary, with much smaller windows.

◄ 0669
Design IV – Landscaping

We believe that landscaping should be indigenous. The garden planting at our own house, Three Seton Mains, is mainly grasses and succulents, plants that are native to the house's coastal location. Species were chosen for growth so the garden was established relatively quickly. We also specifically chose low-maintenance plants, which is an ecological virtue for a busy family.

0670 ▼
Design I – Site layout

Our Cupar houses are carefully located to maximize sunlight and daylight penetration with all primary accommodation facing south. Deep window reveals provide shade in summer and allow sunlight penetration in winter. The project successfully adopts the language and density of the Fife vernacular with environmentally responsive homes designed for lifetime occupation and the changing needs of modern family life.

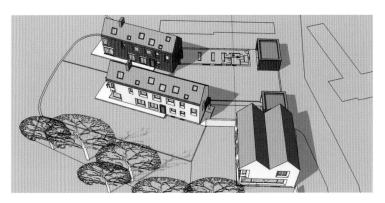

Patrick Nadeau

45 Avenue de la République
Paris, 75011, France
Tel.: +33 6 07 06 38 17
www.patricknadeau.com

0671 ➤

The movable and adjustable garden unit is a large teak and stainless-steel cabinet standing on four legs 3 feet (90 cm) above the ground. When the four sides of the box are opened and placed horizontally, they are transformed into greenhouse tables. Plants grow in the thick trays, drawing liquid nutrients from the channels filled with coconut fiber.

0672 ▼

A mixture of garden and events space, this terrace is distributed into several areas. An esplanade features mobile gardens with a spatial layout that adapts to the life of the company. A small thicket is formed by bushes planted in the ground in leather bags. The main facade reinterprets the Vuitton chessboard with plants and light.

◄ **0673**

Halfway between a garden and an urban square, this space is built using objects and furniture in Ductal, Corian, and stainless steel. Two units resembling bookcases contain the plants. Combined with public benches and Holm oaks in tubs, they form the boundaries of micro-urban spaces designed as many small gardens that intrude on the square.

0675 ▲

A sheet of silvery glass winds its way through a field of red poppies. Out of the presles (aquatic plants) emerge slabs of frosted glass over which float lightly etched motifs. The hydroponic irrigation of the roots appear through the transparent surface. Standing tubular aluminum structures enable plants to be suspended.

◄ **0674**

This 1,400-sq.-foot (130 m²) home nestles under the wave formed by a timber hull. Entirely "greened," the wave-house is a piece of the landscape lightly raised above the ground. It borrows from industrial greenhouse construction in its use of a double polycarbonate skin over a glass facade. The green roof optimizes thermal insulation.

0676 ▼

Plant bags and plant sacks are entirely made of textile. They allow interior spaces to be greened in a number of ways: they can stand on the floor, hang from walls or be suspended from ceilings. Their design responds to the requirements of plant growth. An interior pocket in water-retaining felt and a water reserve make caring for plants easier and reduce water consumption.

◄ **0677**

This installation is made up of epiphytes –(Spanish moss) hanging from structures made in Corian. Spanish moss is native to the forests of South America and the southeastern United States. It lives above the ground, hanging from tree branches and feeding only on the humidity in the air.

0678 ▼

At a workshop, horticulture students met designers and proposed new garden designs. Horticultural techniques, materials, accessories and tools were used to create contemporary forms that combine plants and manufactured articles.

0679 ▼

Between utopia, fiction, and reality, Green Waters embodies a theme of water and plants. The project illustrates the implementation of the latest advances in eco-design in bathrooms (economy, natural capture and filtration of water by plants).

0680 ▼

This project proposes research into the new typologies of objects that incorporate plants. Tables, pots, shelves and green walls offer a view of domestic spaces that is both landscape and landscaped.

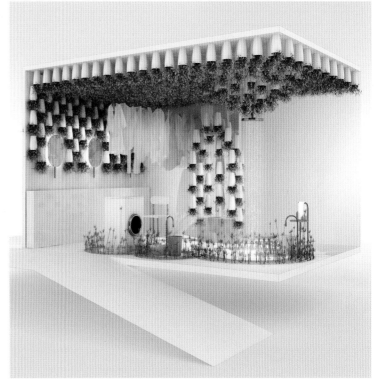

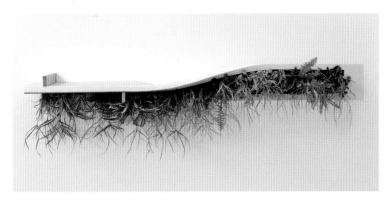

Paul McKean Architecture, LLC.

2505 Southeast 11th Avenue
Portland, Oregon, 97213 USA
Tel.: +1 503 784 3236
www.pmckean.com

0681 ▼
**Reduce the amount of products
in the project by getting double
use from your materials**
In this house we used concrete floors
that act as structural as well as durable
finish flooring. They also provide an
excellent base for radiant floor heating.

0682 ➤
**Separating a building from its
site can have its advantages**
In this case a raised design protects
the house from wild fires and flooding.

0683
Don't under-estimate the 8-foot (2.4 m) ceiling
With the proper window and room proportions, an 8-foot (2.4 m) ceiling can feel tall. This simple move will save on materials, labor and operational expenses.

0684
Pay attention to the common details that get replicated across the entire building
These simple details can raise the quality of design across the whole project.

0685
Have spaces serve multiple purposes for planning efficiency
This canoe loft also serves as an acoustic separation between the main cabin and guest quarters.

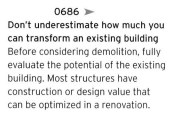

0686 ➤
Don't underestimate how much you can transform an existing building
Before considering demolition, fully evaluate the potential of the existing building. Most structures have construction or design value that can be optimized in a renovation.

0687 ⬈
Design for flexibility
This display wall was built using an adjustable museum fixturing system that allows for infinite variations in the shelving layout.

◄ **0688**
Make it smaller
The best way to decrease the environmental impact of a building is to minimize its footprint. Careful study of use patterns, storage and circulation will lead to more compact designs, and leave more room in the budget for higher quality finishes and systems

◄ 0689

Add to existing structures successfully by having a strong concept for integrating the new and the old

In this master suite renovation we detailed the new construction as temporary cabinetry, providing deep reveals at the connections.

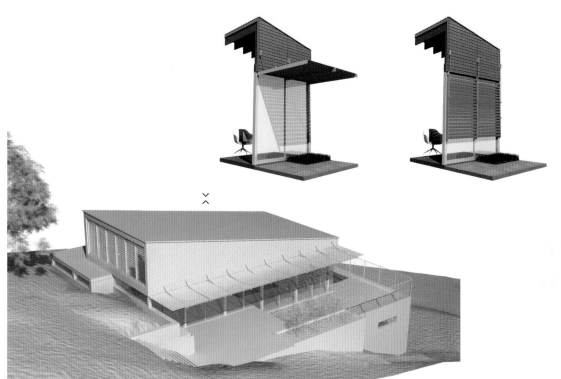

◄ 0690

Use a planning module

Most building products work more efficiently on a 16-inch (40 cm) or 48-inch (122 cm) on-center dimension. Working within these constraints can maximize the abilities of sheathing, formwork, framing and finish panels. This method will also substantially decrease waste.

Paul Morgan Architects

221 Queen Street, Level 10
Melbourne, Victoria, 3000, Australia
Tel.: +61 3 9605 4100
www.paulmorganarchitects.com

0691 ➤
Harness the natural load-bearing capacity of timber found in the region by utilizing bifurcations, (tree forks).

0692 ▼
Get your structural engineer to complete structural optimization testing on the proposed bifurcations, ensuring maximum load-bearing efficiency with the minimum amount of timber.

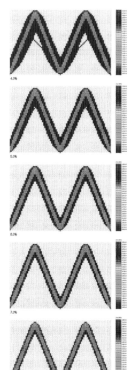

0693 ▲
Ensure the bifurcations are continuous, forming a truss-like frame around the entire house.

0694 ➤
Make sure the structure is continuous, applying internally as well as externally.

0695 ➤
Fix the lining boards, which have moved no more than 164 feet (50 m) from removal to fixing.

0696 ▽
Achieve a six-star energy rating through the use of double glazing, extensive insulation, cement floor slab and overhanging eaves.

0697 ▽
Minimize the carbon footprint. Having removed six trees to make way for the cabin, the trees were milled into lining boards with a mobile milling machine brought on site. The lining boards were seasoned on site.

◁ 0698
Ensure additional lining panels and floor boards are sourced from plantation timber.

0699 ➤
Harvest the rainwater and store it in tanks.

◁ 0700
Investigate the ecological drivers of a region – the study of the relationship between organisms (trees) and their environment has driven the formal/technical design response in this project. Michael Hensel and Achim Menges have coined the term Morpho-Ecology, which relates "a biological paradigm... with issues of higher level functionality and performance capacity."

Peter Kuczia

34 Osningstrasse
Osnabrück, 49082, Germany
Tel.: +49 163 929 50 50
www.kuczia.com

0702 ⬆
The best and least expensive method of
developing sustainably is, of course, not
to build, which is neither satisfactory
for future home owners nor for myself
as architect. The second-best solution
is to create only as much built space
as is really required. Economical
layouts and compact shapes save
resources, energy and money.

0701 ⬇
When possible, reuse existing building
components. Space with old or
vintage construction elements have
a particular spirit and charisma.

0703 ➤

When it comes to chemical-based coverings, a little goes a long way. Sometimes a few accents are all it takes to make a lasting colour impression.

◄ **0704**

Use local, natural materials like wood, for example, in the form of timber cladding for facades.

0705 ▼

Many roofs can be greened. Not only are they good for micro-climate, green roofs insulate as well as protect the roofing materials from UV rays and extreme temperature fluctuations.

◄ 0706
The sun is the most important source of energy on earth. Make use of it in every possible form, for example, for passive heat by creating winter gardens...

0707 ►
... Or by installing solar heating panels for hot water, heating or even for cooling (also possible).

228

0708 ▼
Plenty of natural sunlight saves on energy required for lighting as well as being good for your psyche.

0709 ▲
Loam has a very low embodied energy. Use it for the inside walls. It collects surplus humidity from the air and releases it when the interior is too dry.

◄ **0710**
Reuse building elements. In this way you can conserve resources, reduce waste and save money.

PopovBass Architects

2 Glen Street, Level 3 Milsons Point,
New South Wales, 2061, Australia
Tel.: +61 2 9955 5604
www.popovbass.com.au

"Our aspiration is to leave the world in a better state than when we found it"
Brian Bass on sustainable design

0711 ➤
Architectural language respects the site.

0712 ▼
Light-filled open spaces give joy to their users.

◄ 0713
Economic construction methods, including the use of modular components and prefabricated construction.

◄ 0714
Adapt regional architectural typologies, such as the verandah, to modern planning (this second skin wrapping around the house creates sun protection, privacy and frames views).

0715 ➤
Select sustainable materials
Sustainable materials do not only pertain to recycled and low-carbon products. Sustainable also means low maintenance, robust and timeless, making the building last longer..

0717 ▲
Cross-flow ventilation.

0716 ▼
Facade systems control views, sun and privacy.

0718 ▼
Orientation and creation of overhangs can control natural daylight.

0719 ▲
Low-energy lighting.

0720 ➤
Apply environmental technologies such as solar power, integrated heating systems and rain water collection and recycling.

Prodesi|Domesi

Husitska 36
Prague, 130 00 Czech Republic
Tel.: +420 283 853 424
www.prodesi.cz
www.domesi.cz

0721 ➤
A passive house doesn't have
to look like a science lab.

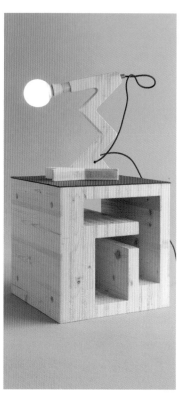

0722 ▲
We use the leftover cuttings from
construction to create furniture.

◄ 0723
We always minimize the
dimensions of a house.

◀ 0724
Our goal is to reduce the impact on the scenery.

◀ 0725
The interior of the house is furnished by local suppliers and finished by a local carpenter.

0726 ➤
To pursue the idea of environmentally conscious architecture, we avoid using exotic woods.

0727 ▼
A house in the countryside should give you the impression it has stood there since time immemorial.

◀ **0728**
A modern wooden house does not have to be just a weekend house in the mountains. Wooden houses are also a part of the confident, clean, urban architecture of today.

◄ 0729
A huge terrace should stand next to the main and compact body of the house – we lose so much heat by biting out the building's bulk.

0730 ►
Combining the traditional, the contemporary and the surroundings should be the way of thinking in modern architecture.

R&Sie(n)

rsie@new-territories.com
www.new-territories.com
http://www.new-territories.com/blog

0731 ▼

Aqua Alta 2.0

A "water bar" located on a Vaporetto
(the waterbus system in Venice,
Italy) serves the French pavilion
contaminated and disgusting
water pumped directly from
the lagoon, sharing substances
and psychological repulsion.

◄ **0732**

Dustyrelief / B_mu

· This structure creates a relief that
was calculated in relation to the
random particles and pixilation
of the pure gray ectoplasm.
Here it is represented against
the light gray Bangkok sky.
· It also collects the city dust
(particles of carbon monoxide)
with an aluminum envelop
and electrostatic system.
· It exacerbates the schizophrenic
environment that separates
the interior (a white cube with
a Euclidian labyrinth) from
the exterior (a dust relief on
topological geometry).

0733 ▼

heShotmeDown

Design of a multi-purpose house,
including private house, dancing center,
shop, restaurant, children's museum...

◄ 0734 ▾

i'mlostinParis

1. This private laboratory was designed to look like a duck blind.
2. It features 1,200 hydroponic ferns.
3. 300 glass beakers are "blowing" components for growing bacterial culture, producing the bacteria *Rhizobium* to increase the percentage of nitrogen without chemical manure or extra light through refraction.
4. It collects rain for watering plants with a mechanical drop-by-drop system.

0735 ▾

Hypnosischamber

Experience an individual hypnosis session in the research and exhibition "I've Heard About":

1. You enter into a "heterotopian" cognitive room.
2. It lets you dive in a "waking dream" filled with vocal information on "I've Heard About" urbanism.
3. Feel yourself as a nerve termination inside the organic and self-determination growing structure.
4. Keep the speculation and experiment alive as a possibility of transformation of your own biotope.

◄ 0736

Hybridmuscle

A work and exhibition space that would generate its own electricity and thus be "unplugged" from the power grid. (Private commission.)

Design of a museum for experimental architecture

The Frac courtyard is covered with gradual glass stick forms, the goal being to produce a bond to cover the existing building, like a potential "body without organs" (an ongoing and never-ending building process) and, inside the glass thickness, there is a path and a labyrinth access. A scattering script is written to develop the aggregation.

0738 ➤

symbiosis'Hood

How to blur the boundaries between two properties, (As a Proudhonism request: property is theft!)

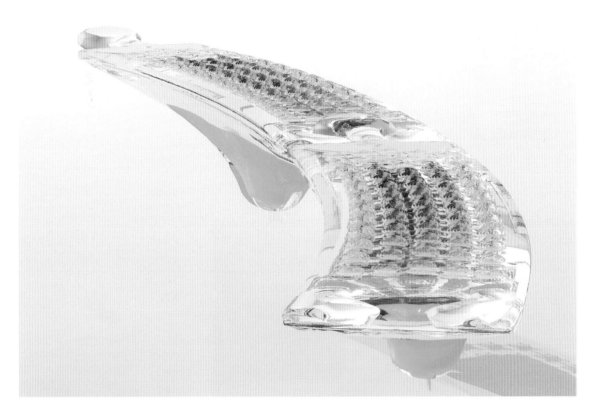

◀ 0739
theBuildingWhichNeverDies
Construction of a light laboratory to test both "the dark adaptation" to reduce urban light pollution and "the melatonin effect" with Oled lamp to test the human metabolism circadian cycle and the UV pathologies from the Ozone weakness evolution.

0740 ▼

waterFlux
1. Digitization of the envelope of a traditional habitat.
2. Scooping out hollows within this volume as if it were an ice cavity, but in full wood with a five-axis drill machine.
3. Water states and flows vary according to the seasons
4. Exacerbation of the winter climate by artificial snow.

Ray Kappe

715 Brooktree Road
Pacific Palisades, California 90272, USA
Tel.: +1 310 459 7791
www.kappedu.com

0741 ➤
Living Homes is a company specializing in prefabricated homes that uses a steel structure, which allows for a multitude of different layouts and finishes.

0742 ▼
The firm seeks to create smart buildings based on passive designer systems, such as a natural heat-monitoring system that harnesses the house's orientation, thermal mass, etc. In addition, it also incorporates some active systems like photovoltaic panels.

◄ 0743
Energy savings of more than 36% of the standard ratings were made thanks to measures that were implemented to reduce the impact on the environment, improve the air quality inside the building, and save energy made through thermal insulation, solar protection, and renewable energy systems, for which the building has earned a LEED Gold rating.

◁ 0744
Associating manufacturing with recycling companies will enable 76% of the materials used in the construction and assembly of the building to be reused when it reaches the end of its useful life.

◁ 0745
The house will therefore be dismantled instead of being demolished, which generates far more waste.

0746 ▷
Amongst its objectives, the firm seeks to reduce the environmental impact, work with natural materials and preserve and reuse natural resources.

0747 ▽
The prefabricated modules have been mounted on top of reinforced concrete blocks of cement, which are good thermal-mass materials.

0748 ▲
Rochedale House is a perfect example of integration between architecture and mass construction. The architectural firm is strongly committed to the ideals of energy efficiency and environmental responsibility.

0749 ▽
Since it is a prefabricated building, the structure of this house can be erected in just 3 days. Apart from this new assembly system, the levels of environmental quality achieved have been recognized by the Leadership in Energy and Environmental Design (LEED) rating system.

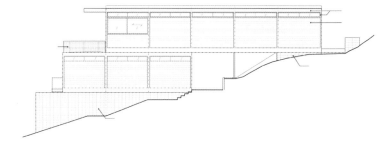

0750 ▷
The sections show the interior spaces of the house and the different levels. The elevations present openings and windows with low-emission glass and certified-wood cladding.

Resolution: 4 Architecture

150 West 28th Street, Suite 1902
New York, New York, 10001, USA
Tel.: +1 212 675 9266
www.re4a.com

0751 ➤
This prefabricated house combination was designed as the best solution for a three-bedroom, two-bathroom house.

0752 ▼
The sustainability of this home is determined by a number of features: the company that manufactured the prefabricated pieces has its own waste-management and material-recycling programs.

◄ **0753**
The photovoltaics panels on the roof and the use of geothermal energy turn the house into a power plant that also supplies the local grid daily.

◄ 0754
The main volume is protected by warm-hued cedar wood while another part of the exterior is covered in cement.

0755 ▲
Prefab construction requires shorter work times, which lessens the environmental impact on the site. Plantings around the site are of native varieties, the lawn is limited, and drought-resistant plants have been used.

0756 ▼
The orientation offers the house the maximum hours of light. Openings are larger on the southern facade, where there is also a large deck and glass-encased porch.

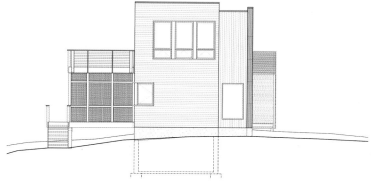

0757 ▲
The kitchen is at the heart of the upper level and separates the living area from the dining area and bedroom. The electricity used comes exclusively from photovoltaic panels and geothermal energy.

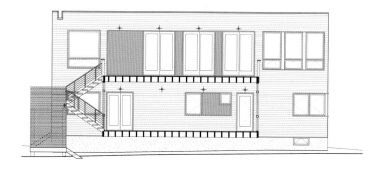

0758 ➤

This prefabricated house is located in a small town in the state of Vermont. Nearby is a state park crossed by the West River and Hamilton Falls. Prefabricated houses involve less impact on the land and lower CO_2 emissions as building times are shorter.

0759
A concrete base was built, and factory-made modules were positioned on it. The house is off-grid; any electricity required is produced by photovoltaic panels.

0760
Radiant floor heating keeps the house warm in the coldest months. Different types of materials were used: recyclable materials, such as the corrugated-meta cladding, and other natural materials, such as cedar, were used on the exterior and the bamboo flooring inside the house.

Rintala Eggertsson Architects

46a Stavangergata 46a
Oslo, 0467, Norway
Tel.: +47 22230006
www.rintalaeggertsson.com

0761 ➤

Narrative

Buildings are not monuments for architects but tools of communication between the inhabitants and the outside world. It is therefore an important part of a design task to communicate a story and connect people with one another. Buildings are like language: they describe our understanding and place in the universe.

0762 ⬥

Physical engagement

Buildings are extensions to our bodies as shields for protection and sensorial membranes. There will always be a physical connection that should activate our bodies and sensitivity toward the environment.

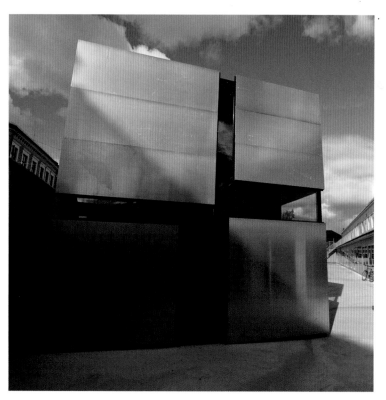

0763 ⬥

Energy savings

Saving building volume is an efficient way to save energy for heating or cooling down the living spaces, which in turn reduces the negative impact on the ecosphere.

⬥ **0764**

Compact living

Compact living is good for the wallet and the environment as it reduces the need for building materials, saves construction time, costs and, consequently, the CO_2 footprint.

0765
Four dimensions

Understanding the environment is to recognize time as the fourth spatial dimension, which gives room for all kinds of changes to a given situation. In an age of global transformation; climatic, demographic, and geo-political changes have to be taken into account.

0766 ➤
Anchoring

Most of our physical environment has been occupied by other human beings in past history, for cultivation, breeding animals or hunting. Preserving a link to these patterns of the past is saving valuable information about the cultural landscape for future generations.

0767 ▲
Site specific

The use of local building materials is an obvious way to establish a visual connection with the surroundings. However, beyond the visual, there is also an important connection to local building traditions that have sustained people in climatically harsh areas around the world for millenniums.

0768
Orientation

In critical climatic situations, directing the building toward a specific cardinal direction can reduce the need for external heating and air-conditioning to a minimum.

0769
Social awareness

Sensitivity toward the physical environment is not the only way to make beautiful buildings. Awareness of social factors is important in order to achieve a well-working relationship between a built structure and the group of people who are going to use it.

0770
Thermal mass

Indigenous people have, throughout history, always taken the physical features of building materials into account when shaping their surroundings. One important feature is thermal mass, which differs from one material to the other and makes it possible to establish a good micro-climate in a building.

Rob Paulus Architects

990 E. 17th Street, Suite 100
Tucson, Arizona, 85719, USA
Tel.: +1 520 624 9805
www.robpaulus.com

0771 ➤
Location efficiency
Location, location, location is key to any successful business but, as it turns out, an urban location also has the advantage of being a more efficient way to live and work. Transit-oriented urban density provides a substantial savings in energy efficiency with the added benefit of direct exposure to the creativity of living within an urban setting.

◄ **0772**
Responsible density
Responsible density is vital to creating walkable, mixed-use cities. Not only is this a good thing for the environment, but it also allows better chances for the financial success of a project while providing amenities and goods within a short distance of home or office.

0773 ▼
Reuse and re-purpose
The most sustainable building is one you don't have to build. Repurposing old structures with new uses gives new life and density to a community without using up as many precious resources. Everyone loves a good story, and authentic historic buildings come with their own tale that can keep evolving and be in step with the new use of the structure.

0774 ➤
Optimism
A healthy dose of optimism is necessary to be able to think clearly to take on creating environments that respond to man and nature. We owe it to our children to be vigilant and create the best possible built environment for their future.

0775 ➤
Make it sexy
An important element of design that isn't talked about enough is the notion of creating something that is both functional and delightful. In other words, creating something sexy that evokes both intellect and intuition. As complex as we humans are, there is also a simple, child-like appreciation and gut reaction to our built environment that defines great design.

◄ 0776
Loose fit
Buildings should be designed with simplicity and adaptability in mind, providing a loose fit that allows for future generations to retrofit and occupy the structures we design today.

0777 ►
Passive first, then active
Building design needs to first acknowledge passive approaches before incorporating active technologies. The basic tenets of natural light and ventilation as well as designing for the region and climate are a solid starting point.

0778 ▼
Healthy habitats
A naturally based environment with day lighting, fresh air and no toxic materials makes us happy and healthy and leads to a more productive lifestyle that is in tune with the daily and yearly cycles of nature.

◄ 0779
Water
Water is the new oil. Almost one-fifth of the world's population, lives in areas of water scarcity. All design should acknowledge water conservation and use this precious resource for inspiration in the creation of buildings, landscapes and cities.

0780 ▲
Humanistic approach
Good design must embrace our existence on this fragile planet we call home and acknowledge our connection to our natural environment to live in balance with Mother Nature.

Rongen Architekten GMBH

2 Propsteigasse
Wassenberg, D 41849, Germany
Tel. +49 2432 3094
www.rongen-architecten.de

0781 ▼

You must build a house that is appropriate for the client. Sustainable architecture is part of nature. Passive house standards are the acknowledged leading standard for energy-saving building. The amount of energy consumption for heating and cooling is around 20% of the amount of a low-energy building, which is the actual standard for new buildings in Germany.

0782 ▲

Sustainable building also means taking the urban environment into account. The design is not just about the details.

 0783
A Passive Houses offers its residents
all the best in living comfort. It's
pleasantly warm in cold winters,
and pleasantly cool in hot summers.
Architecture should always be
clean and clear and thereby true.

0784 ▾
Overhanging floors and balconies
can replace movable sun
protection and thereby further
increase energy efficiency.

0785 ▴
Sustainable building also means
taking nature into account. Passive
Houses save energy resources,
contribute to reducing CO_2 emissions
into the atmosphere and protect
the climate. Our mother planet
does not have any spare parts.

0786 ➤
Photovoltaic panels can replace sun protection while producing green power.

0787 ▼
Sustainable building also means taking the site place into account and building with nature.

0788 ▼
The details and building construction should always be designed in a very neat and tidy manner.

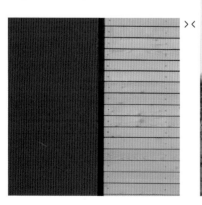

0789 ▽
Sustainable building also means
using renewable building materials.

0790 ▷
Sustainable building also means
space-saving building.

Sustainable Architecture
Advisors International

SAAI has offices in
the UK, Australia, China, Costa Rica,
Thailand, Greece, Mexico and Spain
www.saa-i.com

0791 ▼
Use skylights to improve daylight quality

Compact buildings and deep plans can compromise daylight availability due to the excessive distance from windows to the central core. This is counteracted by the use of skylights, which can provide a more homogeneous daylight distribution, avoiding sharp contrasts and reducing the need for artificial light.

0792 ▶
Transitional spaces

The indoors-to-outdoors transition can be softened with the use of buffer zones that have an undefined character but a great potential for casual usage. In this cultural center located in northwest Spain, a timber structure mitigates the effect of solar radiation during the warm season while allowing useful gains in winter.

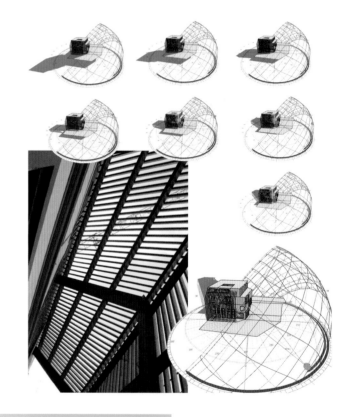

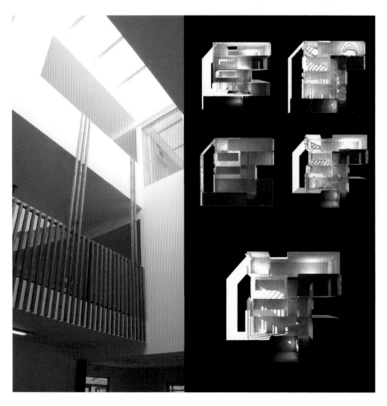

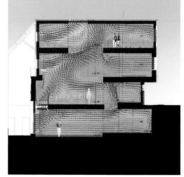

◀ 0793

Buoyancy-driven ventilation is based on the different rate of air pressure reduction at different temperatures. Air pressure decreases with height and, as warm air is less dense, with higher temperatures. This allows incoming (bottom-level) and outgoing (upper-level) air flows to be enhanced when indoor temperatures are warmer than the outdoor temperatures.

0794 ⏶
Materials: Timber
As a natural material, and provided it's responsibly sourced , timber has a relatively low environmental impact. In this cultural center, the potential of timber to provide architectural expression has been explored in various ways. Its different configurations respond to the specific needs of each orientation and the functions behind the facade.

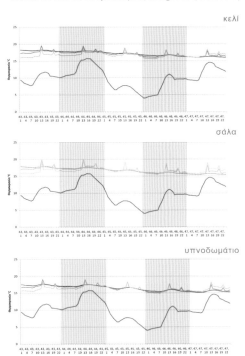

"All year round residence" occupancy scenario simulation results for the three spaces (cell, living room, bedroom)

κελί

σάλα

υπνοδωμάτιο

0795 ⏶
Transitional spaces
In the Museum of Underwater Antiquities of Pylos, Greece, a transitional space allows visual adaptation for the visitors.

◀ 0796
Materials: Stone
Heavy mass stone walls are ideal for the Mediterranean climate, as they ensure stable temperatures. Moreover, stone is an easy-to-find building material in most places, as was the case for this restoration project in Paros, Greece.

Local vernacular
Concepts from the Greek vernacular
architecture have been applied on the
design of this island resort in Paros.
Given its prevailing summer use,
heavy construction is optimal as it
provides stable internal temperature,
including small openings for less solar
gains and visual comfort indoors,
white roofs, shading elements
and a village-like-configuration to
allow ventilation of the resort.

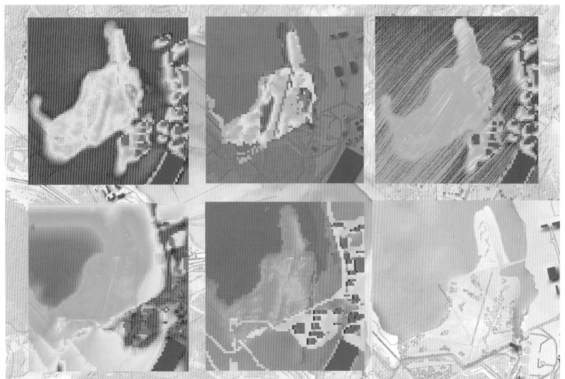

◀ 0798
Sustainable landscapes
Understanding of the elements
that trigger and influence natural
processes is essential to ensure
the correct evolution of the place.
The regeneration of an old quarry
in northwest Spain started with
a detailed study of climatic and
micro-climatic factors, including
prevailing winds, deposition ratio,
thermal flows and soil properties.

0799 ➤

Local vernacular

Elements from the architectural vernacular have also been applied to this housing project in Porto Rafti, Greece. The small size of the openings and the carefully designed shading devices respond to the specific characteristics of the bright daylight in Greece. A projected overhang is a good shading solution for southern orientation in seasonal climates, as it allows sun penetration in winter, when the angle is lower.

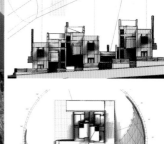

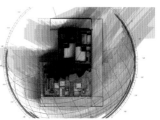

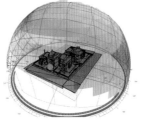

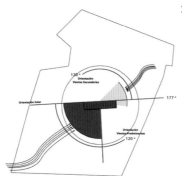

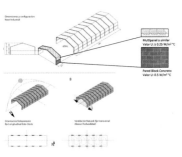

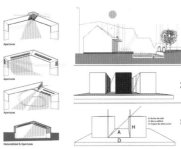

0800 ▲

Wind corridors

In this small community near Bangkok, Thailand, permeability through fragmentation was the key strategy to allow sufficient air movement in and around buildings. In warm and humid climates, ventilation becomes a fundamental strategy to achieve comfort.

Samyn and Partners, Architects & Engineers

Chée/Stwg op Waterloo
Brussels, 1537 B-1180, Belgium
Tel.: +32 2 374 90 60
www.samynandpartners.be

0801 ➤

Consider the sun's path first, its trajectory and rhythms from hour to hour and day to day. Sun penetrates through east and west openings and through the glazed roof to light the cotton fabric that diffuses day light to the architecture classes at the School of Lubumbashi. A termitarium-like structural form manages the natural ventilation and thermal comfort.

0802 ➤

Consider dimensions and quotations. They are the proportion, rhythm and order of magnitude in architecture, just as music theory is to music. Simple volumes and window dimensions impose a rhythm on the campus of the Sans Souci psychiatric hospital.

◄ 0803

Analyze the genius loci. It fosters and tempers the aspirations of the client and feeds your quest. The cultural diversity and unity of Europe is expressed through the patchwork facade of the new European Union Council headquarters, which is made of reused oak windows from all its regions. The project is anchored in an historical building and in the cityscape of Brussels, Belgium.

0804 ▲
Realize that geometry is the most powerful tool in architecture and that drawing is its vector. At Château Cheval Blanc-Saint Emilion, France, the conical stainless-steel tanks in the winery are arranged like organ pipes, and their perforated extensions support the wooden roof shell and filter daylight, while providing safe access to the tank's tops. (preliminary design)

◄ **0805**
Think about space – the voids between solids – as it is the only matter specific to architecture. A large open space for the public is sheltered by the AliBaba headquarter in Shenzhen, China, in contrast to the classical, artificial commercial compounds so common in Asia. (competition entry)

0806 ▲
Study the art of construction: it disciplines the conceptual dream. Invite artists to the game. The former Eastman Institute in Brussels, Belgium is enlarged to house the future museum for European history. The colorful extension by artist Georges Meurant expresses youth and multiculturalism. The exhibition spaces are all naturally lit with three sets of mirrors. (competition entry)

0807 ▲
Accept that matter dictates form. The large wicker huts at the Ngozi cultural center in Burundi are suspended from a knitted aramid net that spans old eucalyptus trees. This allows for large volumes, as their structure is no longer being compressed avoiding buckling. (design stage)

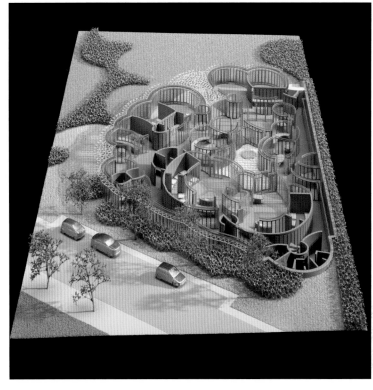

0808 ▼
Ask for lightness, as the art of engineering is to reduce material consumption. The canopy of the Leuven, Belgium train station is designed for ultra-lightness while providing a gently protected space.

0809 ▲
Design lovable spaces – they will be loved. A small daycare for a residential neighborhood in Brussels, Belgium got its rounded form plan from the gentle curved movements the teachers used when explaining the cocoon-like atmosphere they wanted to create. (on hold)

0810 ➤
Dream, think and realize. This meditation octagon is the result of the repetitive unfolding, in two, of the four triangles within the central square.

Sander Architects, LLC

2434 Lincoln, Boulevard
Venice, California, 90291, USA
Tel.: +1 310 822 0300
c@sander-architects.com

0811 ➤
A reflecting pool provides natural cooling. Inside, high windows on the far wall allow rising hot air to escape. This creates a natural convection that sucks the air in from an awning window at pool level: air passing across the pool is naturally cooled, it enters through the low poolside window and passes through the house, lowering temperatures inside.

0812 ▲
Horizontal shade fins protect the south-facing translucent wall from solar infiltration in summer months while allowing plenty of natural daylight. In winter, the lower angle of the sun allows more solar penetration and more natural heating of the space. The wall is constructed with multi-cell translucent material that has twice the R-value of insulated glass and provides excellent insulation.

0813 ▲
Using translucent materials for interior partitions allows light to penetrate to all areas of the house. This way, the house is lit using natural light rather than electric light. Daylight spaces not only use less energy, they have been shown to have a positive effect on mood.

0814 ➤
Eco-friendly recycled blue jean insulation is layered over a thin "space blanket" of foil insulation to double the normal R-value of the walls and roof. This allows the house to be heated or cooled with less energy. In this room, we exposed the blue jean insulation, keeping it in place with a fine wire mesh, so that it doubles as an acoustical layer.

0815 ➤
Recycled steel frames provide the structure for our trademark Hybrid House: part prefab, all custom™ building style that uses warehouse frames as the basic structure and then completes the building with standard construction techniques. This way waste is reduced, costs are lower and construction time is shorter.

◄ 0816
Careful siting of a building allows existing landscape features to remain undisturbed and can provide an interesting design feature. In this case, the external entrance stair to the house is situated to preserve three 100-foot (30.5 m) beech trees on the property. The stairs then cantilever out to capture a view of a seasonal stream before returning to the front door of the house.

0817 ►
This recycled-aluminum staircase is a prime example of using a client's skill set to enhance their home. In this case the client, a lighting designer, had access to a metal fabrication shop used for building lighting grids, and they fabricated the staircase using our design. The recycled metal pulls material out of the waste stream and repurposes it in an interesting way.

0818 ▼
An exterior skin made of recycled steel panels, manufactured with fewer alloys to reduce impact on the environment, is installed in a staggered pattern that echoes the rocks of the distant butte. Time will rust the metal, increasing its resemblance to the butte and avoiding the need for costly maintenance.

0819 ▲
Reusing old structures preserves history and provides an eco-friendly option for creating a new building. In this case, a 300-year-old cotton storage warehouse becomes residences for a group of scientists. Ancient stone walls were all that remained of the warehouse, so the firm inserted modern structure using local materials and recycled steel frames.

0820 ►
The double-height entry staircase in this project captures daylight and fresh air as well as providing a heat chimney for hot air to ascend through the house and escape through vents in the roof. Exposed recycled structural steel creates a sculptural entrance that combines functionality with drama.

Sau, Taller d'Arquitectura S.L.

21-23 Passeig Garcia Faria, Baixos
08005, Barcelona, Spain
Tel.: +34 93 308 6551
www.sausl.com
sautallerarquitectura.blogspot.com

0821 ➤
Can Guetes walkway in Ripoll, Spain
This project touches on collective memory and nature, recovering the historic identity of a site with an industrial past while integrating it into the natural landscape, making it compatible.

0822 ▼
Train depot in Ribes de Freser, Spain
Designing with light, reflections and textures of the environment makes it possible to integrate large volumes into organic and delicate landscapes.

0823 ▼
Colonized fields in Kirchheim, Austria
Architecture as an open system. Architecture should be capable of resolving the current needs using minimum resources. The system can be enlarged or reduced without a trace to meet new needs.

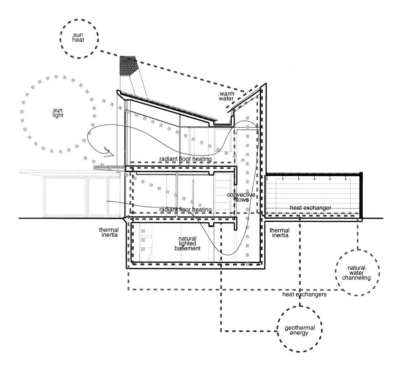

0824 ▲
Plaza del Roser, Sant Joan de les Abadesses, Spain
Minimize physical intervention. At times, making an existing scenario accessible is sufficient; users will handle the rest.

0825 ➤
Cases de l'Estació, Sant Joan de les Abadesses, Spain, with collaboration from architect Agustí Vilà
The city is a changing system in continuous evolution. Town planning, studying and designing a city is about making it a force throughout its evolution, at a social and energetic level.

0826 ⯅
Guarderia La Cabanya, Torelló, Spain
Lights, textures and scale – transfer
the characteristics of the exterior and
filter energy to an interior space to
achieve a spatial sense of the former
and the living conditions of the latter.

0827 ➤
Casa Xemeneia, Torelló, Spain
Integrate architecture, understood as
an energy system, into the environment
and generate exchange flows. Make
use of the existing potential as key to
come closer to energy self-sufficiency.

◄ 0828
Proyecto Mesa de Madera
Having knowledge of the mechanical capacities of the materials and using them transparently is a fundamental tool to optimize resources.

0829 ▼
Guarderia La Cabanya, Torelló, Spain
Permeable architecture, interactive architecture. Architecture should let the environment and light penetrate the interior. Architecture relates with people. Architecture relates people with the landscape.

0830 ➤
Train depot in Ribes de Freser
Reversible systems, such as this energy collectors can be integrated into the landscape during the day. At night, it becomes an architectural landmark that lights the project area by removing the need for additional systems.

Schlyter / Gezelius
Arkitektkontor AB

91, Brännkyrkagatan
Stockholm, 118 23, Sweden
Tel.: +46 708 18 54 35
info@schlytergezelius.se

0831 ▼

Use small local contractors and
carpentry shops. Digital tools were
used to design a handcrafted building,
by updating local building crafts
with today´s digital manufacturing
techniques and using local materials, a
local carpenter and a local contractor.
Use digital tools to calculate the
quantity of building material in order
to create a minimum of waste. All that
was left for me as a keepsake was a
6-inch (150 mm) piece of facade panel.

0832 ▲

Use the conditions of the site with
respect, and make as little impact as
possible. This house was placed for
optimal shelter from the wind and to
obtain the best balance of sunlight to
increase inflow of the sun and decrease
energy losses from the wind. No stone
was blown up and no trees were felled.
Don´t forget the undercurrents of
the water flow. Work with the natural
conditions not against them.

◀ 0833
Integrate with what was there before –
memories in the ground and in people's
minds. In this project, an old stone wall
was incorporated to the building and
was used to equalize heat over time.

0834 ▼
Use green materials. All pieces in
this project were custom made
from wood by a carpenter in the
neighborhood. We were all very
happy when it was finished. It was
a house that smelled good.

0835 ➤
Adjust the shape of the project to
as many site-specific conditions as
possible. This building was shaped
by the terrain and adjusted to rocks,
trees, wind conditions – as many site-
specific conditions as possible. The
forms in the landscape gave rise to
the form. My 3-D software is a great
help when designing site-specific.

0836 ➤

Convert large unused buildings. This old hay loft was converted into an art capsule, a box that was been pushed right through the old building. Two large windows in opposite directions give the studio light. In the interior, white painted hardwood floors and white walls reflect the light. The storage shelves are integrated with the windows. New entrance doors and windows are built into the old facade.

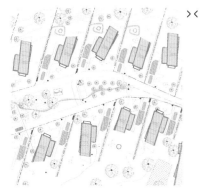 ><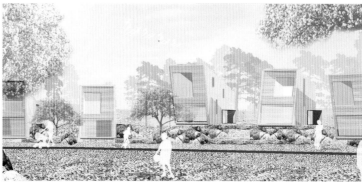

0837 ▼

Details are important. By upgrading details and craftsmanship, we can develop and enrich architecture. Details are the architectural alphabet, our building blocks. Without detail there is no architecture.

 ><

◄ 0838
By reducing surface area and volume we can achieve more architecture. We can use architecture to expand space without expanding the physical space or physical volume.

◄ 0839
Use space to grow edible plants and fruits. This garden was designed to supply fruits and vegetables for one family all year round. The garden contains an open space for cultivation, orchids, an orangery and a greenhouse.

0840 ▼
What kind of society do you want to live in? If we want to understand our global society, it is crucial to understand how our physical landscapes hang together. If we do not understand the physical landscape, how can we ever talk about a sustainable society?

Simon Winstanley Architects

190 King Street
Castle Douglas, DG7 1DB, UK
Tel.: +44 1556 503 826
www.candwarch.co.uk

0841 ➤
The house is located on a spectacular site overlooking the Solway Firth in south-west Scotland. The site is a steeply sloping, former quarry in a National Scenic Area that slopes in two directions from the quarry base, which forms the only level ground access.

0842 ▽
The glazed pavilion is constructed with a steel frame and highly insulated timber infill panels clad in cedar and triple-glazed windows.

◄ **0843**
The external walls, floor and roof are insulated to a high standard and air infiltration is minimized. Triple-glazed windows have warm edge spacer bars and thermally broken frames and inert gas to achieve a very low window U-value.

0844 ➤
The project features a heat pump using a borehole as the ground source for the underfloor heating and hot water system and there is a closed combustion wood burning stove as back up. There is also a micro-generation of renewable electricity using roof mounted photovoltaic panels.

◄ **0845**
Whole-house heat recovery ventilation system.

◄ 0847
The design intention is to create a contemporary single-story "long house" that is recessed in the landscape, sustainable in its construction, very low in energy consumption and aim for zero net emissions of carbon dioxide for all energy use in the house.

0848 ▼
The slope of the roof of the main living accommodation follows the slope of the hillside, with the roof of the rear accommodation meeting the main roof at a shallower angle to allow morning sunlight to penetrate the center of the house.

0846 ▲
The design uses lightweight but highly insulated steel and timber frame construction, clad in cedar weatherboarding, allowed to weather to a natural silver-gray color. The roof uses pre-weathered gray standing seam zinc. Windows and external doors are to be high-performance timber-painted gray.

0849 ⬆
The entrance to the house is sited on the north east-side and under the cover of the roof to provide shelter from the prevailing winds.

><

◄ **0850**
It is proposed to achieve a "zero carbon" house by using very high levels of insulation, minimizing air infiltration, heating using a ground-source heat pump with solar hot water heating panels in the roof and generating electricity using a wind turbine.

SPGArchitects

127 West 26th Street, Suite 800
New York, New York, 10001, USA
Tel.: +1 212 366 5500
www.spgarchitects.com

◄ 0851
Use green roofs
Not only are green roofs aesthetically pleasing, but they help cool the air through evapo-transpiration, the natural process by which water is transferred from plants back into the atmosphere. Additionally, green roofs help provide cleaner air, are great insulators, require less maintenance than conventional roofs and have a longer life than a conventional roof.

0852 ▼
Build with renewable materials
This small house located in Costa Rica is made distinctive by its soaring bamboo structure, a renewable building material that is economical and light-weight. The sculptural roof springs from the ground level, mimicking the rain forest's canopy. This structure filters and shields the interior spaces from the sun and elements while providing views of the rain forest.

◄ 0853
Build as small as possible
This weekend house provides a dramatic rural haven in a very small footprint of 1,076 sq. feet (100 m2). The home's minimal material and resource demands make it a model for efficient living. The interior rooms look toward the views and visually expand the rooms to the landscape beyond.

0854 ▼
Use renewable energy sources
The economics of photovoltaic HIP solar cells improve every year, especially when local tax and power company incentives are included in the calculations. This house is completely powered by solar energy for all of its domestic needs, including lighting, appliances, the plumbing system and swimming pool pumps and other equipment.

◄ 0855
Use pervious paver materials
An alternative to conventional concrete or stone walkways, the pavers used on this entry walk and the green roof allows the rainwater that hits these surfaces to percolate through the material into the ground. In turn, this allows the rainwater to reach the soil without contributing to soil erosion and allows these normally impermeable surfaces to contribute to the refilling of the local aquifer.

0856 ➤

Zone the house to save energy (program zoning)

By using a layout that allows for certain zones to be shut down throughout the year, when they are not needed, the energy demands of this house are decreased as compared to more conventional layouts. This house is heated and cooled in three zones by a geothermal system: the lower level or guest wing, the public living areas, and the master bedroom suite are all independent of one another.

0857 ➤

Use energy efficient lighting

Using a combination of LED (light emitting diode) lighting, along with fluorescent, fiber-optic and low-voltage lighting as well as solar-powered landscape lighting, greatly reduces the energy consumed in this house while still providing a beautiful atmosphere both indoors and out.

0858 ⋀

Use white roof membrane

The highly reflective white membrane roof provides significant energy savings as compared to conventional dark roof surfaces as it reflects the sun's ray and diminishes heat gain associated with the absorption of solar energy. In addition, it does not deteriorate due to exposure to the sun's ultraviolet rays.

◄ 0859
Use low VOC-painting and wood finishes
To protect indoor air quality during the manufacturing process and in the home environment, use paints and wood finishes that are water-based and low in volatile organic compounds (VOC's) to prevent out-gassing of toxic materials from harming the building occupants.

0860 ►
Take advantage of natural ventilation
Maximizing cross-ventilation in each room of this house has eliminated the need for air-conditioning in this year-round tropical climate. Having windows on at least two sides of each room, allows the house to take advantage of the natural convection caused by the cooling of the air as it moves up the hill on which the house sits.

Stuart Tanner Architects

City Mill,
11 Morrison Street
Hobart, 7000, Australia
Tel.: +61 3 6224 4377
www.stuarttanner.com.au

 ><

◄ 0861
Arm End House
Humility and solitude,
protection and openness.

0862 ▼
Goulburn Street Primary
Place and context, linkage and
unification. The new multi-purpose
pavilion provides an alternate entry to
the school, linking two large oak trees
and revealing a broad vista of West
Hobart, Australia, toward the city.

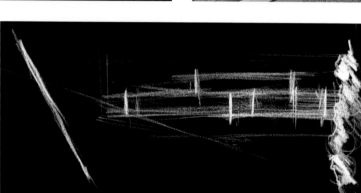  

0863 ▲
The core principle of building less
was fundamental to the project's
sustainable characteristic. A humble
presence on the site also lowered
the impact on the broader context.

0864 ►
Millpond
Shadow and light.

◄ 0865

Millpond
Stillness and memory.

0866 ➤

Not only does this home offer
unfettered visual connections to the
flora and fauna flowing out of the
park, it protects that environment
by minimizing its physical and
environmental footprints.

0867 ➤

Pirates Bay House
Structure and expression,
perception of landscape.

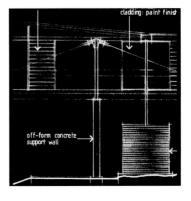

◄ 0868

Pirates Bay Pavilions
Solid and transparent,
interplay of components.

0869 ▲

Passive heating and cooling through
cross-ventilation, on-site waste water
management, rainwater harvesting,
and exterior sun screens are some
of the more impressive architectural
components that make the project
green – in addition to all of those little
things we can all do, such as low-
energy appliances, fixtures, and bulbs.

◄ 0870

Adjacent to the Pirates Bay House
are two small pavilions, nestled into
the terrain, perfect for summer
family use. Both buildings feature
heavy and slender structures as
well as solid and transparent planes,
which open up to the breathtaking
views of the bay below.

Studio 804, Inc / Dan Rockhill, J. L. Constant

Marvin Hall
1465 Jayhawk Boulevard, Room 105
Lawrence, Kansas 66045-761, USA
www.studio804.com

0871 ➤
A narrow footprint carefully positioned within the site maximizes passive strategies and takes advantage of a south face.

0872 ▼
A broad southern facade made up of electrochromic glass – a smart glass that assists in maintaining thermal comfort by mitigating solar heat gain through automated tinting technology – lets the sunlight in during the winter and keeps it out in the summer.

0873 ▲
Rainwater collected from the roof and diverted to an underground cistern helps reduce potable water demands and storm water runoff. The water can be used for exterior landscaping as well as interior uses, such as flushing toilets and irrigating living walls.

◄ 0874
Functioning as storm water detention and a way to decrease urban air temperatures, plant life helps insulate the building, reduces the heat-island effects and helps preserve the roof.

0875 ▼

Helping to conserve natural resources and reducing the load on landfills, a variety of reclaimed/recycled materials can be incorporated into the design. It is also important to be responsible for waste generated on-site. In this example, stone tailings were reclaimed for the facade, and the site waste generated was used along the north side of the building as fill.

0876 ◄

A heavy skeletal frame enables much more insulation to be used in the cavities that exist between the frame materials. Insulating to passive house standards reduces energy loss through the building envelope up to 90%.

0877 ➤

The visibility and accessibility of a building can influence the use of space to publicly demonstrate the latest technologies in design. In this example, a touchscreen panel in the lobby displays real-time energy consumption within the building.

0878 ◄

Most roofing materials are dark in color, but the industry trend to avoid the heat-island effect is to use a light color to reflect heat in climates where air-conditioning in the summer months is necessary.

0879 ➤

Providing priority parking and charging stations for electric vehicles as well as on-site bicycle racks help to encourage the use of alternative transportation.

0880 ▲

Implementing passive strategies of heating and cooling helps to reduce energy loads. One example is a trombe wall, a thermal mass which will absorb the sun's energy during the day and radiate the stored energy at night.

Studio of Environmental Architecture

David Hertz, FAIA, Architects Inc.
Tel.: +1 310 829 9932
www.studioea.com

0881 ➤

Automate
Windows and skylights can be optimally placed for passive natural convective cooling when they are controlled with a simple thermostat and humidistat to become self-regulating. These windows open automatically to the prevailing breezes when the interior reaches a certain temperature and close when it gets too cool or moist out.

><

 0882

Orientation

The McKinley House is situated on the farthest edges of its lot, leaving the center of the property open to the southern exposure and passive solar gain. Sliding doors and windows open the common and private areas to the outside, with threshold-less and continuous floor materials blurring the definition of exterior and interior. The exterior finish provides thermal mass.

0883 ▲

Expose

A large prefabricated monitor was added as an architectural intervention but also serves three primary functions: to provide a flat roof for integrated photovoltaic roofing, to harvest daylight by using the reflective roof and ceiling to bounce light into the space and integrated automated daylight sensors for lighting controls, and to provide for highly operable and thermostatically controlled windows for heat exhaust.

◄ 0884

Simplify

The Panel House uses aluminum clad structurally insulated panels that are typically used for industrial freezers. The high performance panels offer super insulation of R-50, instead of the typical R-19 for walls. Panels are pre-cut, and the building's finished walls are put up in a matter of a few weeks, not months, and require no additional exterior siding, paint or interior finishes, which simplifies materials and minimizes waste.

0885 ▲
Hidden use
The bathroom mirrors in this home slide open to open up the view and reveal or conceal the living space below while providing expansive views from floor-to-ceiling windows and maintaining natural ventilation, where typically a wall would bifurcate the spaces.

0886 ▼
Re- purposing
The 747 Wing House is made from the wings of a retired 747 airplane, which have been re-purposed as its sculptural roof. Use of abandoned or under-utilized structures or materials minimizes the use of raw materials and minimizes the transportation and construction waste.

0887 ▲
Pre-fabricate
This house in Venice Beach uses prefabricated tilt-up concrete walls that were poured off-site and brought to the location. This greatly reduced construction costs compared to poured-in-place walls that would require weeks of building plywood forms or conventional materials that create construction waste. The thermal mass of the walls create an isolation to climate rather than insulation from climate.

><

◄ **0888**

Natural light
Sometimes natural light is reserved only for main areas of the building. Smaller rooms and walkways are often overlooked and often require just as much lighting. This bathroom has an entirely glass ceiling and requires no light during the day. This provides natural light to the halls and rooms during the day as well as cross-ventilation.

0889 ▼

Use the roof
The roof of the Mullin Automotive Museum is easily accessed by patrons via the main elevator and stairs. The roof features a modular green roof of native grasses and plants as well as park-style seating that affords the opportunity to exhibit the solar panels and wind generators. It creates a space with access to the sun, views and ventilation not afforded a typical interior use.

0890 ➤

Use the sun
These parabolic evacuated tube solar collectors capture solar radiation even on cold, cloudy days and take up much less area than typical panels. This system is so effective that it can produce enough hot water for domestic hot water and hydronic radiant heating needs for an entire family.

Studio Negri

Unit 87, The Academy Building, Parkwest Pointe,
Dublin, 12, Ireland
Tel.: +353 86 8212674
www.studionegri.ie

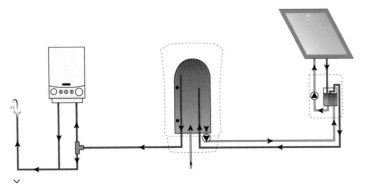

0891 ➤
Create an eco-corridor
Green Connections, a proposed
artists center in Ballymun, Dublin,
Ireland links the newly built structures
into the existing green spaces. All
the trees and vegetation are linear,
and even the green roofs of the
new buildings blend into the public
park. This creates an architectural
eco-corridor and a less fragmented
approach to designing eco-habitats.

0892 ▲
Combi-boilers and solar panels
This house in Dundrum, Dublin, Ireland,
has a system that uses solar panels
in conjunction with combi-boilers.
The solar panel collectors feed into
a buffer tank that directly heat the
showers and hot taps, but if this water
is not hot enough, a thermostatic
valve diverts the pre-heated water
to the combi-boiler, which boosts
it to the required temperature.

0893 ➤
Maximize interaction with nature
This proposal for a social care
center uses high-level windows and
a court-yard in order to maximize
natural light. The narrow plan and
the forming around the court-yard
also allow natural ventillation.

0894 ➤
Internal courtyards maximize natural light
These dark circular walls are both tall and thick, and this maximizes the absorption of solar energy during the day. These Trombe walls have vents added to the top and bottom of the interior wall, which allow heated air to flow via convection into the building interior.

0895 ▼
Roof garden with water retention tanks
This roof garden, on a house in Seafield, in South Africa, not only allows optimized views and additional outdoor living space, it also provides resistance to thermal radiation and reduces rain runoff. By connecting the runoff to water-retention tanks, this water is reused in the garden and connected to flushing toilets.

0896 ▼
Low-maintenance and recycled materials
This attic conversion uses both zinc and recycled timber. Zinc is 100% recyclable, very long lasting and does not require maintenance. Recycled timber used internally can give a much better effect than a new timber floor. Always analyze the embodied energy of a material; try to find the sum total of the energy necessary for the product's entire lifecycle.

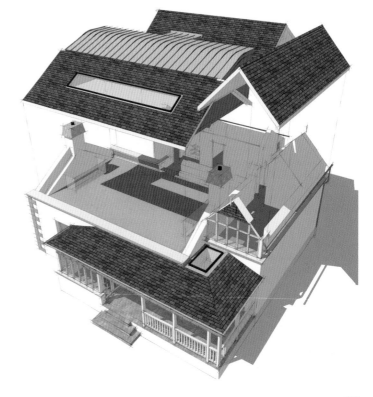

◄ 0897
Optimize south orientation with generous overhangs
This overhang is a shading device to reduce glare and net gain during the day and heat loss at night. For overhangs to be effective they must be carefully designed to be appropriate to the location and orientation.

0898 ▼
Natural light
Electric lighting contributes to internal heat gains in buildings, which reduces the building's winter heating load, but increases the summer cooling load. Introduce surfaces of high reflectance and maximize natural daylight, thereby reducing artificial lighting and associated costs. Natural light also allows internal planting to thrive as much as possible.

◄ 0899
Natural ventilation
The rate of ventilation required by the occupants of a space must be fully understood prior to designing. Natural ventilation works best in narrow spaces, and 39 feet (12 m) wide or less works better.

0900 ➤

Masterplan showing combined principles

As well as normal drawings, and architectural masterplans, also try to explain by diagram all the eco-mechanisms and strategies. This is a proposal for a sports hall in Dalkey in Dublin, Ireland. It contains a multitude of eco-solutions: solar energy, geo-thermal ducts, heat recovery systems and rain-water harvesting.

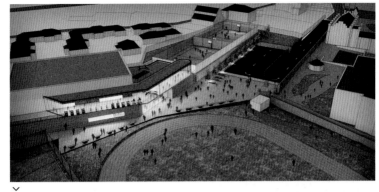

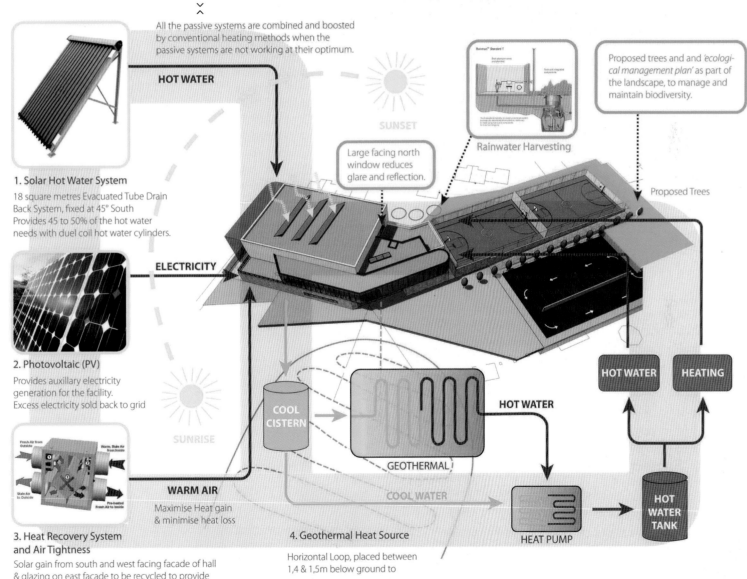

All the passive systems are combined and boosted by conventional heating methods when the passive systems are not working at their optimum.

HOT WATER

SUNSET

Proposed trees and and *'ecological management plan'* as part of the landscape, to manage and maintain biodiversity.

Rainwater Harvesting

Large facing north window reduces glare and reflection.

1. Solar Hot Water System

18 square metres Evacuated Tube Drain Back System, fixed at 45° South Provides 45 to 50% of the hot water needs with duel coil hot water cylinders.

Proposed Trees

ELECTRICITY

2. Photovoltaic (PV)

Provides auxillary electricity generation for the facility. Excess electricity sold back to grid

SUNRISE

COOL CISTERN

GEOTHERMAL

HOT WATER

HOT WATER   HEATING

COOL WATER

HOT WATER TANK

WARM AIR

Maximise Heat gain & minimise heat loss

HEAT PUMP

3. Heat Recovery System and Air Tightness

Solar gain from south and west facing facade of hall & glazing on east facade to be recycled to provide

4. Geothermal Heat Source

Horizontal Loop, placed between 1,4 & 1,5m below ground to

Suppose Design Office

15-1 Funairihon-machi
Naka-ku, Hiroshima
730-0843, Japan
Tel. : +81 82 961 3000
www.suppose.jp

0901 ➤
Being situated on a small lot between two buildings, the house receives no natural light from normal windows. The architects put almost all of the windows on the roof.

0902 ▼
It was our intention to treat rooms and gardens as equivalent, and make the relationship between inside and out closer, by creating a design that features this garden-like room so that things normally decorating a room, such as art, books and furnishings, would, in a way, be thrust into an almost exterior space.

0903 ▲
Using this design as a starting point, we hope that words such as garden and landscape, which have only been used for exteriors, can begin to take on new and varied meanings, bringing vibrant and beautiful scenery into the interior of homes as well, and make architectural aesthetics more and more diverse.

◄ **0904**
Bright colors, large breakthroughs in walls, transitions without any thresholds and a garden room that extends through the entire house and, by means of skylights, provides the rooms with natural light.

◄ **0905**
Rather than a design that begins to grow stale as soon as it is completed, through this design featuring the constantly changing and vibrant "garden room," we hope that the tenants' daily lives will be richer than before.

0906 ➤
It is now recognized that indoor plants are able to cleanse the air and improve the chemical composition of interior environments.

0907 ▽
The upper floors are composed only by the pyramid-shaped roof. The main feature of the house is perhaps its skylight, which connects each floor.

0908 ▽
The site, a former field, has been incorporated into the design for bearing stratum. The soil leftover from excavating the ground was used to make the hill, which functions as the garden on the exterior and provides privacy.

0909 ▲
The banisterless central staircase has a dual function: it connects the house's different levels, and it forms a volume that occupies the core space.

◀ **0910**
The surplus land from the earthwork was reused to shape the hill that acts as an outdoor garden.

Taalman Koch Architecture

1570 La Baig Ave, Unit A
Los Angeles, California, 90028, USA
Tel.: +1 213 380 1060
www.taalmankoch.com

0911 ➤

Less is less

Less clutter, less noise, less distraction, less energy, less material, less mess. Less is about stealth design and minimal spaces, letting the occupants, objects and landscape interact with the space. Strip back to the bare structural elements, leave the floor bare as exposed concrete, leave the ceiling exposed and define the space with the minimum of material.

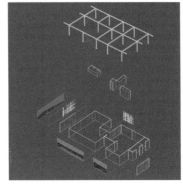

◄ 0912

Prefabricate elements and chunks or even the whole building

Fabricating and finishing parts off-site is more precise and efficient than on-site labor and is less wasteful. We minimize construction waste by pre-cutting, drilling and finishing the elements of the itHouse of-site.

◀ 0913
Smaller and smarter parts = less shipping and less on-site work
Construction can be messy, often requiring large equipment and temporary access. By breaking the parts of this prefabricated house down to small parts, shipping is optimized and sites can be minimally disturbed. No cranes, no temporary roads, all the parts can be brought in by small crews and assembled with few tools.

0914 ▼
Bring the indoors out and outdoors in
Unify the inside with the outside by bringing exterior materials indoors and opening up large views and openings to the exterior. Use materials to connect the indoor and outdoor spaces.

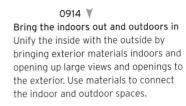

0915 ▲
Integrate the systems seamlessly
Buildings and spaces today must interface with lighting and heating, cooling systems to create, power and control their environments. We believe in seamlessly integrating technology into the design and allowing those systems to either be completely hidden from view or to become design elements. The solar panels for the off-grid itHouse double as a shade canopy for the exterior courtyards.

0916 ⋀
Use simple natural materials, use few materials
Keep the palette to one or two materials if possible. Use color and wallpaper to articulate space as an inexpensive way to have a big impact. We use durable and natural floor materials like concrete and cork that can perform well in all types of spaces to create a unified plane.

0917 ➤
Reuse, recycle and upcycle
Reuse and recycle materials found on-site, or upcycle those materials for new uses. Cut up the old slab and make a stacked concrete wall or remove old siding and create a new fence. New materials can be created from old. In the Purdy Devis House we repurposed materials that needed to be removed from the old house and used them as landscape elements.

0918 ➤

Work with what you have
Adaptively re-using old structures makes use of existing building stock and can create new hybrid spaces and programs. The existing conditions are often taken for granted and thrown away.

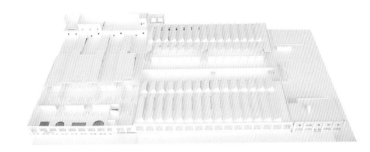

0919 ▼

Maximize and optimize spaces by creating multi-functional elements
By doubling on the function of a space, new experiences emerge. In the Burns Gorman House we are creating a steel stair whose treads and risers are the structure and form a bookcase, compressing the structural, circulation and storage requirements into a compact and sculptural element.

0920 ▼

Use daylight as much as possible
Daylighting of spaces saves energy and allows the space to change as the natural light fluctuates. Natural light offers the best light quality to view color and detail. As the light changes the space changes, allowing the interior to reflect the exterior mood.

Theis and Khan Architects, Ltd.

22 Bateman's Row
London, EC2A 3HH, UK
Tel.: +44 20 7729 9329
www.theisandkhan.com

0921 ➤

Density: Bateman's Row
By densifying an area with this mixed-use project, the urban fabric is stitched back together. A more sustainable community can be established, one that does not rely on cars to get around. Enabling people to live in the city can create safer environments. This efficient alternative to sprawling development allows rural areas to be maintained.

0922 ▼

Solar energy: Bateman's Row
Harnessing the readily available energy of the sun reduces the dependency on gas and electricity in this mixed-use project. These solar panels produce the majority of hot water for the building.

◄ 0923

Green roof: Bateman's Row
A variety of planting is integrated into this inner-city project to enhance the biodiversity of the area, provide solar shading and improve the privacy of the family areas. By reducing hard surfaces, water run off can be dealt with without overloading the public system.

◄ 0924

Reuse: Lumen URC
An existing church is reconfigured and extended rather than rebuilt to give it a new lease on life. The structure is highly insulated to reduce heat loss and features efficient new services that minimize energy use and running costs. Community rooms and a café are incorporated to provide a valuable resource within walking distance for the local community.

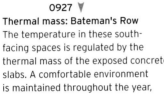

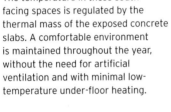

◄ 0925
Facade design: Bateman's Row
The detailed design of facades and selection of materials ensure the building is as efficient as possible. The proportion of glazed to solid areas is carefully considered to provide optimal thermal, ventilation, light and aesthetic results.

◄ 0926
Natural ventilation: Lumen URC
A former parking garage is transformed into a garden, providing herbs and vegetables for the community café, a quiet sanctuary for locals, and encouraging biodiversity. Community rooms are naturally ventilated with a combination of low-level vents in the glazing and opening rooflights at the rear of the room, which also provide natural light deep into the plan.

0927 ▼
Thermal mass: Bateman's Row
The temperature in these south-facing spaces is regulated by the thermal mass of the exposed concrete slabs. A comfortable environment is maintained throughout the year, without the need for artificial ventilation and with minimal low-temperature under-floor heating.

0928 ▲
Off-site construction: Lumen URC
Extensions are constructed from cross-laminated timber panels as a truly sustainable alternative to traditional materials. As well as having the lowest energy consumption of any building material across its lifecycle, the off-site production approach reduced the overall building program and project cost while increasing precision and build quality.

◄ 0929
Materials: Friends House
In line with the clients' Quaker values, the approach for this project had sustainability at its core, using simple, economic materials from renewable sources, such as the spruce plywood for all of the joinery.

0930 ▲
Lighting design: Lumen URC
Natural light is celebrated and combined with artificial lighting to allow a variety of uses and atmospheres in the space. Concealed low-energy linear lighting is both economic and elegant. By switching the up and down lighting separately, as well as controlling each bay, flexibility of light levels and character is integrated with minimum fixtures and energy use.

TYIN Tegnestue Architects

51 Kjøpmanssgata
Trondheim, 7011, Norway
Tel.: +47 452 55 502
www.tyintegnestue.no

◄ 0931
An existing boathouse from the 1800s became not only the mold for the new structure, but parts of the materials from the old structure were used as cladding in the new. The old, rusty, corrugated steel sheets became beautiful doors in the new facade.

0932 ▼
A crucial part of reuse is the aspect of adding value. Old, dusty timber can become a valuable asset for communities where resources are scarce, but only when some precision and care is put into the work. A dab of fresh paint over some old ply-wood boxes accentuates the new and sets this wall apart from any other wall.

0933 ▼
In hot, humid areas where electricity is limited and expensive, natural ventilation is an obvious choice. Here, the large roof covers the heavy brick structures, keeping the courtyard and spaces colder during the day and warmer at night. Woven mats of bamboo prevent radiation of heat from the corrugated metal roof.

0934
When working in the rural areas of Thailand, transporting materials is a big obstacle. The natural choice was to use locally grown bamboo found near the building site. The material is free, grows back rapidly and is in tune with local traditions and skills.

0935
Involving people from the local community is crucial for ensuring that a project is socially viable. Through conversations, drawing workshops and model-making, people of all ages can join in the discussion.

0936
During the process of creating architecture, surprises will arise. When the bedrock was exposed under mud and grass, the best choice was to incorporate the beautiful surface in the space itself. Being aware of opportunity during all the parts of the process usually strengthens the project.

0937
Strategical use of durable materials like concrete and steel can improve the longevity of any design. The low bench gives a place to rest after the ball game while also strengthening the existing brick wall.

0938
Old windows found near the building site were used to create the openings in the back facade. The size of the frames determined the main axis for the structure. Combined with new construction timber treated with eco-friendly oils, the old material gets a new value.

0939
Find simple solutions to basic challenges. By combining plastic piping, ready-made concrete tubes and locally harvested timber, an open-air bathhouse is made for Tasanee and one of her orphaned children.

0940
Involvement creates a sense of ownership and pride in the people involved.

UCARCHITECT

283 Lisgar Street
Toronto, Ontario, M6J 3H1, Canada
Tel.: +1 416 536 4977
www.ucarchitect.ca

◄ 0941
**Site sensitivity and storm
water management**
Existing heavily treed ravine forest,
contours that rise 85 feet (26 m) in
height and drainage patterns influence
the building's proportions, orientation
and location. The building is embedded
into the hillside, and, despite its
large floor plate, blends in with its
surroundings due to its extensive
green roof vegetation and its one-to
two-story appearance from three sides.

0942 ➤
**Improvement of existing
contextual condition**
Located within a suburban context and
yet physically surrounded by heavy
vegetation on nearly all sides, the
building's natural finishes and calm
colors integrate it with its setting.
Translucent glazing at the entrance
allows for natural light to enter and
yet affords privacy. The building is a
pleasure for the eye from many angles.

The edge separating the inside from the outside is intentionally blurred with tall glass sliding doors. The easily accessible outdoor decks encourage outdoor living in a country known for long winters and utilize otherwise unused roof areas.

0944 ▲
Passive daylighting
Large glazed openings and skylights make artificial lighting unnecessary during the day. Natural daylight penetrates all spaces deep within the building. South-facing elevations provide optimal daylight while interior and exterior shading devices offer additional solar control.

◄ 0945
Highly insulated exterior walls, roof, windows, and doors
The building uses wood, both as a structural and finishing element, to complement the austerity of the light floors and walls.

◄ 0946
Land use efficiency
Half perched, half nestled into its site, the building extends across the site with a light footprint, fully visible from the narrow access road but hidden from the lakeside. A generously sized wrap-around deck creates a close link between the interior and exterior.

0947 ▼
Passive ventilation
Interconnected spaces spanning all floors allow for natural cross ventilation by creating unimpeded air flow through the building. Open views are maintained throughout.

◄ 0948
Radiant floor heating embedded within pigmented poured concrete. Thermal mass provides even temperature comfort throughout all seasons.

0949 ➤

Compact and yet spacious

A modest building with a floor plate of only 1238 sq. feet (115 m²), yet spacious due to its enlarged angled volume, interconnected spaces, eliminated hallways, dual use of spaces, and well-placed views and vistas.

> <

0950 ⋏

Low environmental impact over the building's life cycle

The exterior is composed of durable weather-resistant and maintenance-free materials: unfinished red western cedar and sheets of light-coloured metal. The interior and furnishings are void of unnecessary accents, baseboards, or mouldings. Basic elements and varied textures take precedence.

UdA, Architetti Associati

68 Via Valprato
Torino, 10155, Italy
Tel.: +39 112 489489 int. 201
www.uda.it

0951 ▼
Coherence lies in never hiding the past. Better to change it into a challenge for innovation.

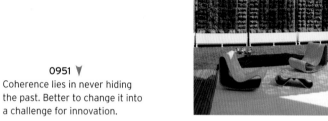

0952 ▼
Look upon workplaces, be they factories or offices, as they are often scenes wherein the best of human talent is applied. For this reason, they must always be accompanied by nature and its elements: light, greenery, air.

0953 ▲
Never lose touch with the land, particularly when this may become a refined, aesthetic enjoyment and a contact with life.

0954 ▲
Surround yourself with lighthearted things of which you are fond, bearing in mind that their simplicity will always be a source of pleasure.

0955 ➤
Sustainability of a building may be understood in many ways. We like to think that this term can also embrace the ability of architecture to converse with its context, with coherence, but also everyone in their own way.

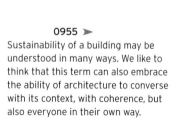

0956 ▼
Use the more unusual or commonplace spaces and street furniture as instruments of a consciousness of our senses, bodies and everyday actions.

0957
In certain contexts, respect for a locality and its traditions must not take the form of mindless imitation of ancient forms. What is required is innovation with embodiment of the spirit of things and the traces of human endeavours.

0958 ➤
Go back to a certain friendship spirit, not at the beck and call of fashion or some ephemeral trend, but as an aid in rediscovering the human component that lives in houses, offices and factories. A coffee and snack corner in an office may become a new place for a meeting with colleagues.

0959
A city must be the recipient of gentle grafts that serve to upgrade its existing buildings. New parts embodying better technologies improve the performance of already built-up spaces.

0960 ⅄
Think of benches, seats and other facilities provided in public spaces as places of emotional experience for physical and psychological well-being (something children do not need to be told!).

Vincent Snyder Architects

1711 Briar Street
Austin, Texas 78704, USA
Tel.: +1 512 228 9245
snyder@mail.utexas.edu

0961 ➤
Place windows to induce natural ventilation.

0962 ▽
Control heat gain during the summertime and heat loss during the winter.

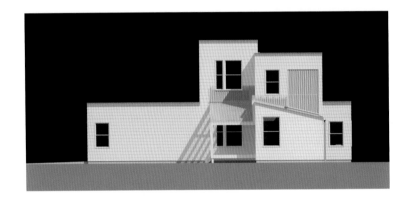

0963 ➤
Low-VOC paints are less toxic for the environment.

0964 ▽
Low-maintenance durable cladding gives buildings a longer life, whether using an expensive material like slate or an inexpensive material like corrugated metal.

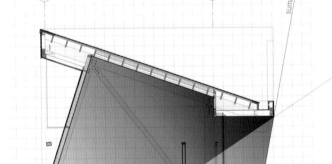

 0965 ➤
Shaded openings on the northern and southern facades bring in light and views but diminish direct sun exposure.

0966 ▼
Use concrete slab for thermal mass.

0967 ▲
Use concrete slab for thermal mass.

0968 ➤
Solid massing on eastern and western facades mitigates early morning and late afternoon sun intrusion.

0969 ➤
Create covered exterior spaces that double as buffer zones for indoor heat gain.

0970 ▼
Use light-colored finishes for reflection to lower heat gain.

WD Architects

30 King Street
Cooran, Queensland 4569, Australia
Tel.: +61 485 2720
www.wdarchitects.com.au

0971 ➤
Land use and ecology
Projects are designed as an adaptation
to their immediate environment.
Working predominantly in the
subtropics, the climate is wet and
humid during the summer with a short
cold, dry winter and a mild spring
and autumn. The pattern of living
in this region favours the outdoors,
with the connection between indoors
and outdoors always important.

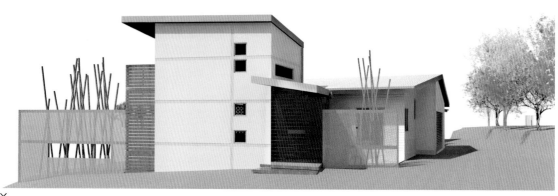

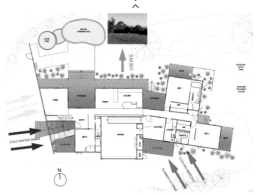

◄ 0972
Water
Australia is one of the driest continents
on the planet, so every project collects
all its own water for internal and
landscape use. Waste water is either
for internal reuse or recycled for
landscape use. Hot water is heated
via solar systems mounted on the
roof, and all plumbing fixtures and
appliances are selected for their
high energy and water efficiency.

◄ 0973

Materials

Materials are chosen for their low toxicity, emissions and embodied energy, high level of insulation or thermal capacity. They are also selected for their low impact on ecosystems and biodiversity when grown, harvested or manufactured. Their ability to be reused, recycled or composted at the end of the life of the building is another consideration, as are local materials where they are available.

0974 ▼

Energy

Buildings are designed to be low users of energy and to heat and cool themselves, reducing and eliminating mechanical systems. Solar orientation, passive heating and cooling, cross- and stack-ventilation, and insulation are fundamental methods of reducing energy consumption employed in all WD Architects projects as shown in the design of the Maleny Sports Club.

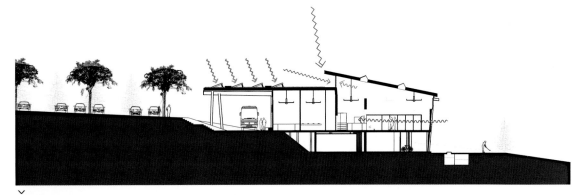

0975 ➤

Air quality

Natural ventilation is an essential part of every design for cooling and air quality. The Noosa Pengari Steiner School Hall uses cross and stack ventilation, as well as a large-diameter low-speed fan to move air through the building. Air is also drawn across the lawn and from garden plants nearby adding further cooling and filtering effects to maintain a consistently high internal air quality regardless of the weather conditions.

0976 ▼

Emissions

Designing for less energy and water consumption means less emissions into the environment and atmosphere. Composting toilets are used at the Noosa Pengari Steiner School to minimize water consumption and use composted waste for re-vegetating the school site. Minimizing the light emission from a project is another important issue to reduce the impact on local ecology at night.

0977 ▼

Size

Size affects everything, so building less means less cost and impact on the environment. Buildings are conceived with less material, and materials are finished in their raw state where possible. Smaller buildings mean less energy and water consumed in their construction and over their life span. Internal space is made to feel bigger by including the outside as part of the inside and extending the landscape into the building.

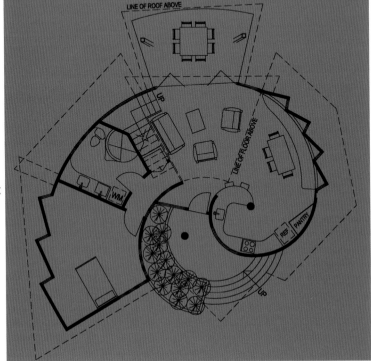

0978 ▲

Innovation

Every project is a new investigation of how to do things better, often using innovative strategies and technologies that achieve improved sustainable outcomes. The mainstays though are time-honored low-tech methods and solutions that allow buildings to regulate themselves through passive means with only fine tuning required by the occupants for individual comfort levels.

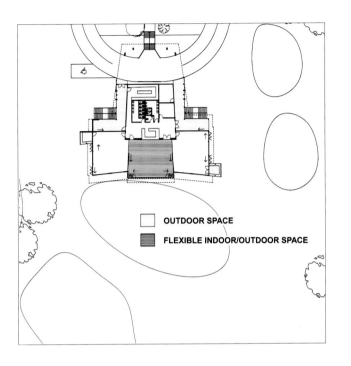

OUTDOOR SPACE

FLEXIBLE INDOOR/OUTDOOR SPACE

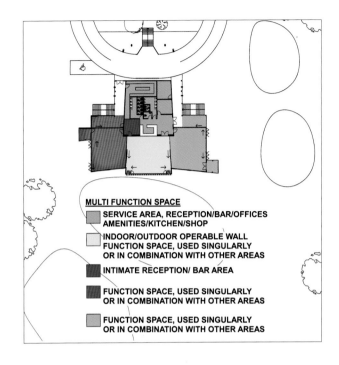

N

MULTI FUNCTION SPACE

SERVICE AREA, RECEPTION/BAR/OFFICES
AMENITIES/KITCHEN/SHOP

INDOOR/OUTDOOR OPERABLE WALL
FUNCTION SPACE, USED SINGULARLY
OR IN COMBINATION WITH OTHER AREAS

INTIMATE RECEPTION/ BAR AREA

FUNCTION SPACE, USED SINGULARLY
OR IN COMBINATION WITH OTHER AREAS

FUNCTION SPACE, USED SINGULARLY
OR IN COMBINATION WITH OTHER AREAS

0979 ⬆
Life cycle
The complete life cycle of a project is designed from its inception. Not only the cost of construction but the cost of operating the building needs to be considered, along with flexibility for future changed use, reuse-ability and recyclability at the end of its life span.

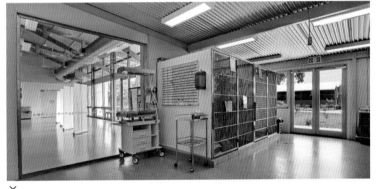

◀ 0980
Education and community
There is an education component to every project WD Architects delivers. They demonstrate the effectiveness and beauty of sustainable design and give a greater understanding of the issues and possibilities. This is especially true in a school context, such as the Noosa Pengari Steiner School, and in the very public Wildlife Warriors Veterinary Hospital.

Werner Sobek Stuttgart GmbH & Co. KG

14 Albstr.
Stuttgart, 70597, Germany
Tel.: +49 711 76750 38
www.wernersobek.com

◄ 0981
The sum is greater than the sum of its parts
Holistic design and interdisciplinary planning visions are developed in-house by a tightly knit team of experts who develop both individual and universal strategies and systems.

◄ 0982
Triple zero®: Zero energy
Our proof-of-concept buildings generate all energy required during operation on-site.

0983 ▼
Proof of concept
Our buildings demonstrate state-of-the-art research and design. They are prototypes of building solutions for sustainable architecture in the 21st century.

0984 ➤
Triple zero®: Zero emissions
Energy is generated exclusively by renewable energy sources; no fossil fuels are used on-site.

0985 ➤
Continuous exploration
Research is an integral part of our work and design. New materials and technology are incorporated when they support the overall goals of the design. Simulations support the design process, and monitoring can verify assumptions and create the base for future predictions.

0986 ➤
Design for disassembly
The end-of-life phase of a building is taken into account at all stages of the design and building process.

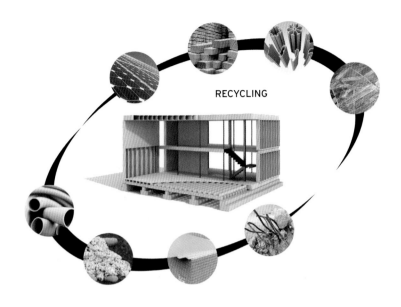

RECYCLING

0987 ▼
Beauty
Essential and deeply human, breathtaking beauty creates affection for and a deep bond with our built environment.

0988 ⋀
Triple zero®: Zero waste
Our buildings are fully recyclable; thus, they don't deplete natural resources but supply valuable resources for future generations.

0989 ▼
Urbanism
Individual buildings form the fabric of our built environment and can incorporate alternative solutions to urban issues – for example, by integrating E-mobility into the architecture and technology of the house.

0990 ➤
High tech at its best: Cutting-edge technology
The latest technology and software employed for the design and operation processes of our buildings. Technology is fully integrated into the design, but it is never a means on its own.

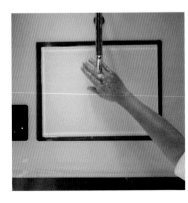

ZeroEnergy Design

156 Milk Street, Suite 3,
Boston, Massachusetts, 02109, USA
Tel.: +1 617 720 5002
www.zeroenergy.com

0991 ➤

Take your shoes off!

Your shoes track in chemicals and contaminants that you don't want in your healthy home. Create a mudroom or mud area near the entry door so it's convenient for family and visitors to remove shoes when entering your house.

0992 ▼

Borrowed light

Interior glazing can transfer light into spaces that don't have direct access to daylight, minimizing the need for artificial light.

◀ **0993**

Thermal mass is often over rated. Check with your energy professional before adding unnecessary concrete to your floor assembly or using a high-mass/reduced R-value wall system. Unless you live in a climate with big daily temperature swings (like the desert), a simple tile will probably provide the benefit you're looking for.

◄ 0994
Prioritize south-facing glazing
In cold climates in the northern hemisphere, high performance south-facing windows will offer a net gain over the heating season. The passive solar heat gained through the windows will exceed the heat lost. East-, west-, and north-facing windows, however, will be a net loss.

0995 ▼
Scenarios for living
In planning a new home, think about how you will use each space on a day-to-day basis and for the unique scenarios it will accommodate, like Thanksgiving dinner for 20. By planning for multifunctional space, you can minimize square footage – the easiest way to conserve cost, energy and material resources.

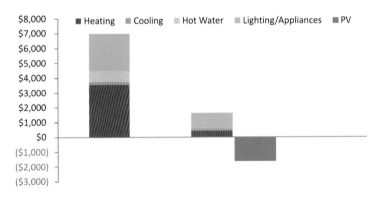

0996 ▲
Crunch the numbers and design the mechanical system
Work with a professional who is knowledgeable in the field of energy modeling and mechanical design to ensure that your house meets both your energy performance and thermal comfort expectations. Not sure where to find one? A Certified Passive House Consultant should have the right skill set.

0997 ➤

Avoid carpet

Carpet will trap dust, dirt, mold spores, and other allergens and contaminants. Solid-surface flooring is easy to clean and a great fit for radiant heat. As a budget-friendly and unique alternative to stained hardwood, consider a more economical wood floor (or even a plywood subfloor) paired will a no-VOC floor paint.

0998 ▼

Mechanical ventilation with energy recovery

Build your home as airtight as possible. Incorporate whole-house mechanical ventilation with an energy or heat recovery ventilator (ERV or HRV) to maintain excellent indoor air quality. A whole-house HEPA filter can also be added to reduce common allergens.

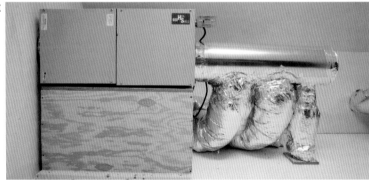

◄ 0999

Plan for solar

Even if it's not in the budget on day one, if your site allows, plan for solar. Decide if you'd like to see the panels on the roof or conceal them. By making the requisite roof penetrations during the initial home construction, you can properly flash them and avoid future leaks, which are often a result of cutting holes in your roof after the fact.

> <

1000 ►

Indirect lighting

Using an indirect light source is a great way to light a space. Warm up a cool light source, like a linear fluorescent lamp, by bouncing light off a wood or plywood ceiling. The thin aluminum bar in this photo is a fluorescent light fixture. An alternative detail is to create a cove with a more economical fluorescent strip light.